Addition and Subtraction

Combining and Comparing

Grade 3
Also appropriate for Grade 4

Jan Mokros
Susan Jo Russell

Developed at TERC, Cambridge, Massachusetts

Dale Seymour Publications®
Menlo Park, California

The *Investigations* curriculum was developed at TERC (formerly Technical Education Research Centers) in collaboration with Kent State University and the State University of New York at Buffalo. The work was supported in part by National Science Foundation Grant No. ESI-9050210. TERC is a nonprofit company working to improve mathematics and science education. TERC is located at 2067 Massachusetts Avenue, Cambridge, MA 02140.

This project was supported, in part,
by the
National Science Foundation
Opinions expressed are those of the authors
and not necessarily those of the Foundation

Managing Editor: Catherine Anderson
Series Editor: Beverly Cory
Revision Team: Laura Marshall Alavosus, Ellen Harding, Patty Green Holubar, Suzanne Knott, Beverly Hersh Lozoff
ESL Consultant: Nancy Sokol Green
Production/Manufacturing Director: Janet Yearian
Production/Manufacturing Coordinator: Barbara Atmore
Design Manager: Jeff Kelly
Design: Don Taka
Illustrations: DJ Simison, Carl Yoshihara
Cover: Bay Graphics
Composition: Archetype Book Composition

This book is published by Dale Seymour Publications®, an imprint of Addison Wesley Longman, Inc.

Dale Seymour Publications
2725 Sand Hill Road
Menlo Park, CA 94025
Customer Service: 800-872-1100

Order number DS43847
ISBN 1-57232-700-6
1 2 3 4 5 6 7 8 9 10-ML-01 00 99 98 97

 Printed on Recycled Paper

TERC

Principal Investigator　Susan Jo Russell

Co-Principal Investigator　Cornelia C. Tierney

Director of Research and Evaluation　Jan Mokros

Curriculum Development

Joan Akers
Michael T. Battista
Mary Berle-Carman
Douglas H. Clements
Karen Economopoulos
Ricardo Nemirovsky
Andee Rubin
Susan Jo Russell
Cornelia C. Tierney
Amy Shulman Weinberg

Evaluation and Assessment

Mary Berle-Carman
Abouali Farmanfarmaian
Jan Mokros
Mark Ogonowski
Amy Shulman Weinberg
Tracey Wright
Lisa Yaffee

Teacher Support

Rebecca B. Corwin
Karen Economopoulos
Tracey Wright
Lisa Yaffee

Technology Development

Michael T. Battista
Douglas H. Clements
Julie Sarama Meredith
Andee Rubin

Video Production

David A. Smith

*Cooperating Classrooms
for This Unit*

Corrine Varon
Virginia M. Micciche
Cambridge Public Schools
Cambridge, MA

Jeanne Wall
Barbara Walsh
Arlington Public Schools
Arlington, MA

Katie Bloomfield
Robert A. Dihlmann
Shutesbury Elementary
Shutesbury, MA

Consultants and Advisors

Elizabeth Badger
Deborah Lowenberg Ball
Marilyn Burns
Ann Grady
Joanne M. Gurry
James J. Kaput
Steven Leinwand
Mary M. Lindquist
David S. Moore
John Olive
Leslie P. Steffe
Peter Sullivan
Grayson Wheatley
Virginia Woolley
Anne Zarinnia

Administration and Production

Amy Catlin
Amy Taber

Graduate Assistants

Kent State University

Joanne Caniglia
Pam DeLong
Carol King

State University of New York at Buffalo

Rosa Gonzalez
Sue McMillen
Julie Sarama Meredith
Sudha Swaminathan

Revisions and Home Materials

Cathy Miles Grant
Marlene Kliman
Margaret McGaffigan
Megan Murray
Kim O'Neil
Andee Rubin
Susan Jo Russell
Lisa Seyferth
Myriam Steinback
Judy Storeygard
Anna Suarez
Cornelia Tierney
Carol Walker
Tracey Wright

CONTENTS

WHERE TO START

The first-time user of *Combining and Comparing* should read the following:

When you next teach this same unit, you can begin to read more of the background. Each time you present the unit, you will learn more about how your students understand the mathematical ideas.

Investigations in Number, Data, and Space® is a K–5 mathematics curriculum with four major goals:

■ to offer students meaningful mathematical problems

■ to emphasize depth in mathematical thinking rather than superficial exposure to a series of fragmented topics

■ to communicate mathematics content and pedagogy to teachers

■ to substantially expand the pool of mathematically literate students

The *Investigations* curriculum embodies a new approach based on years of research about how children learn mathematics. Each grade level consists of a set of separate units, each offering 2–8 weeks of work. These units of study are presented through investigations that involve students in the exploration of major mathematical ideas.

Approaching the mathematics content through investigations helps students develop flexibility and confidence in approaching problems, fluency in using mathematical skills and tools to solve problems, and proficiency in evaluating their solutions. Students also build a repertoire of ways to communicate about their mathematical thinking, while their enjoyment and appreciation of mathematics grows.

The investigations are carefully designed to invite all students into mathematics—girls and boys, members of diverse cultural, ethnic, and language groups, and students with different strengths and interests. Problem contexts often call on students to share experiences from their family, culture, or community. The curriculum eliminates barriers—such as work in isolation from peers, or emphasis on speed and memorization—that exclude some students from participating successfully in mathematics. The following aspects of the curriculum ensure that all students are included in significant mathematics learning:

■ Students spend time exploring problems in depth.

■ They find more than one solution to many of the problems they work on.

■ They invent their own strategies and approaches, rather than relying on memorized procedures.

■ They choose from a variety of concrete materials and appropriate technology, including calculators, as a natural part of their everyday mathematical work.

■ They express their mathematical thinking through drawing, writing, and talking.

■ They work in a variety of groupings—as a whole class, individually, in pairs, and in small groups.

■ They move around the classroom as they explore the mathematics in their environment and talk with their peers.

While reading and other language activities are typically given a great deal of time and emphasis in elementary classrooms, mathematics often does not get the time it needs. If students are to experience mathematics in depth, they must have enough time to become engaged in real mathematical problems. We believe that a minimum of five hours of mathematics classroom time a week—about an hour a day—is critical at the elementary level. The plan and pacing of the *Investigations* curriculum is based on that belief.

We explain more about the pedagogy and principles that underlie these investigations in Teacher Notes throughout the units. For correlations of the curriculum to the NCTM Standards and further help in using this research-based program for teaching mathematics, see the following books:

■ *Implementing the* Investigations in Number, Data, and Space® *Curriculum*

■ *Beyond Arithmetic: Changing Mathematics in the Elementary Classroom* by Jan Mokros, Susan Jo Russell, and Karen Economopoulos

This book is one of the curriculum units for *Investigations in Number, Data, and Space.* In addition to providing part of a complete mathematics curriculum for your students, this unit offers information to support your own professional development. You, the teacher, are the person who will make this curriculum come alive in the classroom; the book for each unit is your main support system.

Although the curriculum does not include student textbooks, reproducible sheets for student work are provided in the unit and are also available as Student Activity Booklets. Students work actively with objects and experiences in their own environment and with a variety of manipulative materials and technology, rather than with a book of instruction and problems. We strongly recommend use of the overhead projector as a way to present problems, to focus group discussion, and to help students share ideas and strategies.

Ultimately, every teacher will use these investigations in ways that make sense for his or her particular style, the particular group of students, and the constraints and supports of a particular school environment. Each unit offers information and guidance for a wide variety of situations, drawn from our collaborations with many teachers and students over many years. Our goal in this book is to help you, a professional educator, implement this curriculum in a way that will give all your students access to mathematical power.

Investigation Format

The opening two pages of each investigation help you get ready for the work that follows.

What Happens This gives a synopsis of each session or block of sessions.

Mathematical Emphasis This lists the most important ideas and processes students will encounter in this investigation.

What to Plan Ahead of Time These lists alert you to materials to gather, sheets to duplicate, transparencies to make, and anything else you need to do before starting.

INVESTIGATION 1

Comparisons with Record Numbers

What Happens

Sessions 1 and 2: How Many Children in Your Family? Students compare their own data on the number of children in a family with the record number of children listed in *The Guinness Book of Records.* Students are also introduced to Close to 100, a game they will play throughout the unit, in which they compose a number that is close to 100 and compare this number with 100.

Session 3: More Record Comparisons Students bring in data from home and compare it with world record data. They compare the record ages of animals and people with the ages of their own pets and relatives.

Mathematical Emphasis

■ Comparing two numbers and developing strategies for figuring out the difference between them

■ Developing ways of getting close to 100 by combining numbers

■ Using landmark numbers (multiples of 10 and 100) to help in making a comparison of two quantities

---

What to Plan Ahead of Time

Materials

■ Interlocking cubes or counters (along with the 100 charts and 300 charts you will be duplicating) should be available to all students throughout the unit for use in modeling problems.

■ *The Guinness Book of Records, 1993* compiled by Peter Matthews (Bantam Books, 1993): 1 or 2 copies (Sessions 1–3, optional). A new edition is published every year and your students might have fun finding the latest issue, but the statistics we use here have not changed for many years.

■ Numeral Cards: 1 deck per 2-3 students (Session 1). If you do not have manufactured sets, make your own; see Other Preparation.

Other Preparation

■ Duplicate student sheets and teaching resources (located at the end of this unit) in the following quantities. If you have Student Activity Booklets, copy only the item marked with an asterisk.

For Sessions 1–2

100 chart (p. 81): 1 per student

Numeral Cards (p. 103): enough to make one deck per 2 or 3 students (see below)

How to Play Close to 100 (p. 106): 1 to post, or 1 per student, optional

Student Sheet 1, Close to 100 Score Sheet (p. 79): 1 per student

Student Sheet 2, Ages (p. 80): 1 per student (homework)

Family letter* (p. 78): 1 per student. Remember to sign it before copying.

For Session 3

300 chart (p. 107): 1 per student

■ If you haven't purchased the *Investigations* grade 3 materials kit, make a deck of Numeral Cards for every two or three students. If you can duplicate the sheets on tagboard, the cards will last longer. Cut apart the 44 cards for one complete deck. If students help with cutting, many sets can be made quickly. Mark the back of each deck differently, or copy the sheets for each deck on paper of a different color, so that the decks are easy to separate if they get mixed up in use. (Sessions 1–2)

■ Play the Close to 100 game by yourself or with someone else, so that you are familiar with the rules and scoring. (Sessions 1–2)

■ If you plan to provide folders in which students will save their work for the entire unit, prepare these for distribution during Sessions 1 and 2.

■ If you have pets in your classroom or school, find out how old they are. (Session 3)

■ There is interesting background material about centenarians in *The Guinness Book of Records*, including information on countries with the highest rates of longevity. You may want to share this with students. (Session 3)

Sessions Within an investigation, the activities are organized by class session, a session being at least a one-hour math class. Sessions are numbered consecutively through an investigation. Often several sessions are grouped together, presenting a block of activities with a single major focus.

When you find a block of sessions presented together—for example, Sessions 1, 2, and 3—read through the entire block first to understand the overall flow and sequence of the activities. Make some preliminary decisions about how you will divide the activities into three sessions for your class, based on what you know about your students. You may need to modify your initial plans as you progress through the activities, and you may want to make notes in the margins of the pages as reminders for the next time you use the unit.

Be sure to read the Session Follow-Up section at the end of the session block to see what homework assignments and extensions are suggested as you make your initial plans.

While you may be used to a curriculum that tells you exactly what each class session should cover, we have found that the teacher is in a better position to make these decisions. Each unit is flexible and may be handled somewhat differently by every teacher. While we provide guidance for how many sessions a particular group of activities is likely to need, we want you to be active in determining an appropriate pace and the best transition points for your class. It is not unusual for a teacher to spend more or less time than is proposed for the activities.

Ten-Minute Math At the beginning of some sessions, you will find Ten-Minute Math activities. These are designed to be used in tandem with the investigations, but not during the math hour. Rather, we hope you will do them whenever you have a spare 10 minutes—maybe before lunch or recess, or at the end of the day.

Ten-Minute Math offers practice in key concepts, but not always those being covered in the unit. For example, in a unit on using data, Ten-Minute Math might revisit geometric activities done earlier in the year. Complete directions for the suggested activities are included at the end of each unit.

Sessions 1 and 2

How Many Children in Your Family?

Materials

- Interlocking cubes
- 100 charts
- Numeral Cards (1 deck per 2–3 students)
- Student Sheet 1 (1 per student)
- How to Play Close to 100 (1 to post, or 1 per student, optional)
- Student Sheet 2 (1 per student, homework)
- Family letter (1 per student)
- *The Guinness Book of Records* (1 or 2 copies, optional)

What Happens

Students compare their own data on the number of children in a family with the record number of children listed in *The Guinness Book of Records*. Students are also introduced to Close to 100, a game they will play throughout the unit, in which they compose a number that is close to 100 and compare this number with 100. Student work focuses on:

- developing strategies for comparing two numbers
- exploring combinations of numbers that make 100
- using landmark numbers while making comparisons

Activity

Comparing Family Size

Explain to the class that they will be making many comparisons during this unit. One of the things they will do is make comparisons with *The Guinness Book of Records*:

Today you are going to compare the number of children in your family to the record number of children in a family. How should we count the number of children in our families? Whom should we count?

Students—particularly those with step-siblings and blended families—will probably have questions about which people to count. Defining how to count children in the family is an important mathematical task. Allow some time to discuss and decide who should be counted and help students arrive at a consensus. Questions like these may come up:

I'm a foster child. Which family should I count?
My mom's pregnant. Can I count the new baby?
My mom has a new husband and he has two kids that live with us. Do I count them?
My dad has a kid from his old wife. Should I count her?
My sister's married. Does she count?

4 ■ *Investigation 1: Comparisons with Record Numbers*

Activities The activities include pair and small-group work, individual tasks, and whole-class discussions. In any case, students are seated together, talking and sharing ideas during all work times. Students most often work cooperatively, although each student may record work individually.

Choice Time In some units, some sessions are structured with activity choices. In these cases, students may work simultaneously on different activities focused on the same mathematical ideas. Students choose which activities they want to do, and they cycle through them.

You will need to decide how to set up and introduce these activities and how to let students make their choices. Some teachers present them as station activities, in different parts of the room. Some list the choices on the board as reminders or have students keep their own lists.

Extensions Sometimes in Session Follow-Up, you will find suggested extension activities. These are opportunities for some or all students to explore a topic in greater depth or in a different context. They are not designed for "fast" students; mathematics is a multifaceted discipline, and different students will want to go further in different investigations. Look for and encourage the sparks of interest and enthusiasm you see in your students, and use the extensions to help them pursue these interests.

Excursions Some of the *Investigations* units include excursions—blocks of activities that could be omitted without harming the integrity of the unit. This is one way of dealing with the great depth and variety of elementary mathematics— much more than a class has time to explore in any one year. Excursions give you the flexibility to make different choices from year to year, doing the excursion in one unit this time, and next year trying another excursion.

Tips for the Linguistically Diverse Classroom At strategic points in each unit, you will find concrete suggestions for simple modifications of the teaching strategies to encourage the participation of all students. Many of these tips offer alternative ways to elicit critical thinking from students at varying levels of English proficiency, as well as from other students who find it difficult to verbalize their thinking.

The tips are supported by suggestions for specific vocabulary work to help ensure that all students can participate fully in the investigations. The Preview for the Linguistically Diverse Classroom (p. I-22) lists important words that are assumed as part of the working vocabulary of the unit. Second-language learners will need to become familiar with these words in order to understand the problems and activities they will be doing. These terms can be incorporated into students' second-language work before or during the unit. Activities that can be used to present the words are found in the appendix, Vocabulary Support for Second-Language Learners (p. 75). In addition, ideas for making connections to students' language and cultures, included on the Preview page, help the class explore the unit's concepts from a multicultural perspective.

Materials

A complete list of the materials needed for teaching this unit is found on p. I-18. Some of these materials are available in kits for the *Investigations* curriculum. Individual items can also be purchased from school supply dealers.

Classroom Materials In an active mathematics classroom, certain basic materials should be available at all times: interlocking cubes, pencils, unlined paper, graph paper, calculators, things to count with, and measuring tools. Some activities in this curriculum require scissors and glue sticks or tape. Stick-on notes and large paper are also useful materials throughout.

So that students can independently get what they need at any time, they should know where these materials are kept, how they are stored, and how they are to be returned to the storage area. For example, interlocking cubes are best stored in towers of ten; then, whatever the activity, they should be returned to storage in groups of ten at the end of the hour. You'll find that establishing such routines at the beginning of the year is well worth the time and effort.

Technology Calculators are used throughout *Investigations*. Many of the units recommend that you have at least one calculator for each pair. You will find calculator activities, plus Teacher Notes discussing this important mathematical tool, in an early unit at each grade level. It is assumed that calculators will be readily available for student use.

Computer activities at grade 3 use two software programs that were developed especially for the *Investigations* curriculum. *Tumbling Tetrominoes* is introduced in the 2-D Geometry unit, *Flips, Turns, and Area.* This game emphasizes ideas about area and about geometric motions (slides, flips, and turns). The program *Geo-Logo*™ is introduced in a second 2-D Geometry unit, *Turtle Paths,* where students use it to explore geometric shapes.

How you use the computer activities depends on the number of computers you have available. Suggestions are offered in the geometry units for how to organize different types of computer environments.

Children's Literature Each unit offers a list of suggested children's literature (p. I-18) that can be used to support the mathematical ideas in the unit. Sometimes an activity is based on a specific children's book, with suggestions for substitutions where practical. While such activities can be adapted and taught without the book, the literature offers a rich introduction and should be used whenever possible.

Student Sheets and Teaching Resources Student recording sheets and other teaching tools needed for both class and homework are provided as reproducible blackline masters at the end of each unit. They are also available as Student Activity Booklets. These booklets contain all the sheets each student will need for individual work, freeing you from extensive copying (although you may need or want to copy the occasional teaching resource on transparency film or card stock, or make extra copies of a student sheet).

We think it's important that students find their own ways of organizing and recording their work. They need to learn how to explain their thinking with both drawings and written words, and how to organize their results so someone else

can understand them. For this reason, we deliberately do not provide student sheets for every activity. Regardless of the form in which students do their work, we recommend that they keep a mathematics notebook or folder so that their work is always available for reference.

Homework In *Investigations,* homework is an extension of classroom work. Sometimes it offers review and practice of work done in class, sometimes preparation for upcoming activities, and sometimes numerical practice that revisits work in earlier units. Homework plays a role both in supporting students' learning and in helping inform families about the ways in which students in this curriculum work with mathematical ideas.

Depending on your school's homework policies and your own judgment, you may want to assign more homework than is suggested in the units. For this purpose you might use the practice pages, included as blackline masters at the end of this unit, to give students additional work with numbers.

Name _____ Date _____

Student Sheet 4

Who Is Closer to 100?

Two students are playing Close to 100.

Student 1 has   Student 2 has

6 6 1 8 3 5   3 9 2 7 5 8

Find the 4 cards that will get each player as close as possible to 100.

Student 1   Student 2

☐☐ + ☐☐ = _____   ☐☐ + ☐☐ = _____

Who got closer to 100? _____

Choose one student's hand. Explain why this is as close as you can get to 100 with those six cards.

_____
_____
_____
_____
_____
_____
_____

© Dale Seymour Publications®   83   *Investigation 2 • Session 2*
                                                  Combining and Comparing

For some homework assignments, you will want to adapt the activity to meet the needs of a variety of students in your class: those with special needs, those ready for more challenge, and second-language learners. You might change the numbers in a problem, make the activity more or less complex, or go through a sample activity with those who need extra help. You can modify any student sheet for either homework or class use. In particular, making numbers in a problem smaller or larger can make the same basic activity appropriate for a wider range of students.

Another issue to consider is how to handle the homework that students bring back to class—how to recognize the work they have done at home without spending too much time on it. Some teachers hold a short group discussion of different approaches to the assignment; others ask students to share and discuss their work with a neighbor, or post the homework around the room and give students time to tour it briefly. If you want to keep track of homework students bring in, be sure it ends up in a designated place.

Investigations at Home It is a good idea to make your policy on homework explicit to both students and their families when you begin teaching with *Investigations*. How frequently will you be assigning homework? When do you expect homework to be completed and brought back to school? What are your goals in assigning homework? How independent should families expect their children to be? What should the parent's or guardian's role be? The more explicit you can be about your expectations, the better the homework experience will be for everyone.

Investigations at Home (a booklet available separately for each unit, to send home with students) gives you a way to communicate with families about the work students are doing in class. This booklet includes a brief description of every session, a list of the mathematics content emphasized in each investigation, and a discussion of each homework assignment to help families more effectively support their children. Whether or not you are using the *Investigations* at Home booklets, we expect you to make your own choices about home-

work assignments. Feel free to omit any and to add extra ones you think are appropriate.

Family Letter A letter that you can send home to students' families is included with the blackline masters for each unit. Families need to be informed about the mathematics work in your classroom; they should be encouraged to participate in and support their children's work. A reminder to send home the letter for each unit appears in one of the early investigations. These letters are also available separately in Spanish, Vietnamese, Cantonese, Hmong, and Cambodian.

Help for You, the Teacher

Because we believe strongly that a new curriculum must help teachers think in new ways about mathematics and about their students' mathematical thinking processes, we have included a great deal of material to help you learn more about both.

About the Mathematics in This Unit This introductory section (p. I-19) summarizes the critical information about the mathematics you will be teaching. It describes the unit's central mathematical ideas and how students will encounter them through the unit's activities.

Teacher Notes These reference notes provide practical information about the mathematics you are teaching and about our experience with how students learn. Many of the notes were written in response to actual questions from teachers, or to discuss important things we saw happening in the field-test classrooms. Some teachers like to read them all before starting the unit, then review them as they come up in particular investigations.

Dialogue Boxes Sample dialogues demonstrate how students typically express their mathematical ideas, what issues and confusions arise in their thinking, and how some teachers have guided class discussions.

These dialogues are based on the extensive classroom testing of this curriculum; many are word-for-word transcriptions of recorded class discussions. They are not always easy reading; sometimes it may take some effort to unravel what the students are trying to say. But this is the value of these dialogues; they offer good clues to how your students may develop and express their approaches and strategies, helping you prepare for your own class discussions.

Where to Start You may not have time to read everything the first time you use this unit. As a first-time user, you will likely focus on understanding the activities and working them out with your students. Read completely through each investigation before starting to present it. Also read those sections listed in the Contents under the heading Where to Start (p. vi).

Using Concrete Materials for Addition and Subtraction
> Teacher Note

Many of your students will need to model problems in some way that helps them see more clearly the problem situation and what they are trying to find out. Help students use cubes, counters, 100 charts, or 300 charts to show you how they are thinking about the problem. The following examples show how students might use these materials:

Cubes or Counters Suppose a student is comparing a family of 4 children to the 69 children in the "record" family. This student might build 69 with cubes—making 6 rows of ten and 1 row of nine, then put 4 connected cubes on top of the 69, then count how many more children there are in the group of 69.

Another student might see this problem as finding the difference between 69 and 4, and so might build 69 with the cubes, then take 4 away, and count how many are left. A third student might just imagine the 60, build 9, take away 4 cubes from the 9, and then mentally add back the 60.

As students are counting with cubes, gradually help them move beyond counting by 1's. Some students cling to counting by 1's because they feel this is the only way they can be sure of their accuracy. However, counting by 1's is difficult, inefficient, and prone to error—for anyone. Help students build confidence and skill in counting by 2's, 5's, and 10's. If you keep your interlocking cubes stored in towers of 10, students will be able to work with them more easily.

100 Chart and 300 Chart Using the 100 chart to make the same comparison (the difference between 4 children and 69 children), students could place a counter or mark on 4 on the chart, and then find how many steps or jumps it will take to reach 69. At first, students may count by 1's. Encourage them to figure out how to make bigger jumps, such as from 4 to 10, then to 20, 30, 40, 50, 60, and finally to 69. Demonstrate how to jot down intermediate differences while jumping along the chart (6, 10, 10, 10, 10, 9).

One of the biggest problems students have on the 100 chart is understanding where to start counting. When finding the difference between 69 and 4, for example, many students count both the 4 and the 69, ending up with a difference of 66 instead of 65. Help these students by working with much smaller numbers:

How many more children would you need in your family to make 5? 6? 7? Why don't you say 5 is two bigger than 4?

Students who persist with counting the beginning number should return to using cubes or counters. Reformulate the problem in this way:

Here are the 4 children in your family. I'll make them with red cubes. I want you to add on blue cubes until you have 69. Then we can see how many extra children you'd need to have 69 in your family.

If some students are finding differences correctly on the 100 chart but are counting by 1's, help them begin to take bigger jumps by asking questions like these:

Where would you land if you took 10 steps? Now what if you took 10 more steps?

As students work with problems with larger numbers, they will find that the 300 chart can replace the 100 chart to help them visualize the relationships among the numbers.

D I A L O G U E   B O X

How Do You Count the Days?

The following discussion took place in a classroom that was working on the first calendar activity, How Much Longer? (p. 60).

Everybody put an X on today, the 12th. OK. What's happening next Tuesday?

Michael: We're going on our field trip.

Right—that's the 17th. How many days from today is that?

Various students: Five . . . Six . . . Four.

How did you get your numbers?

Su-Mei: It's five because I counted from each day to the next day. From today till tomorrow is 1, then till Saturday is 2, till Sunday is 3, till Monday is 4, and till Tuesday is 5.

Annie: But today's a day, so we should count it. We should count the 12th, 13th, 14th, 15th, 16th, and 17th. That's six days.

Rashad: It's five days because today's not a full day. It already started.

Well, it's 9:30 now. Suppose we were starting our trip at 9:30 on Tuesday?

Yoshi: That would be four days.

How did you get that?

Yoshi: Tomorrow, Saturday, Sunday, Monday. So, four days.

Hmmm. So we have a lot of disagreement. Some people think we should count the day we start on. Some people don't. Some people think we should count from one day to the next as one day. What if your birthday were on Saturday? How many days away would you say it is?

Tamara: I'd say two, because there's almost all of today to get through, and then all of tomorrow, and then you wake up and it's your birthday.

Jeremy: Yes, it's two, but my reason is different. I'd say from today till tomorrow is one day, and

from tomorrow till Saturday is one day, so it's two days away.

Liliana: No, it's just one. I'd say it's one day away because today has already started and so you just count Friday.

Su-Mei: But then what if it was tomorrow? You don't say tomorrow is zero days away.

It's an interesting question, and one not easily answered. Most adults have decided that the way to count days is to count from one day to the next. So that tomorrow would be one day away, Saturday two days away, and Sunday three days away. If we use that method, how long is it from now until our field trip?

Mark: Five days. It really is closest to five days.

Annie: I still say it's six. It is if you count my way!

This conversation highlights an important mathematical issue—whether you are counting objects or intervals. These students are grappling with an important question: do you count days or the amount of time that elapses from one day to another? In this conversation, students are also beginning to think about what is discrete (such as individual objects) and what is continuous (such as measurements).

The *Investigations* curriculum incorporates the use of two forms of technology in the classroom: calculators and computers. Calculators are assumed to be standard classroom materials, available for student use in any unit. Computers are explicitly linked to one or more units at each grade level; they are used with the unit on 2-D geometry at each grade, as well as with some of the units on measuring, data, and changes.

Using Calculators

In this curriculum, calculators are considered tools for doing mathematics, similar to pattern blocks or interlocking cubes. Just as with other tools, students must learn both *how* to use calculators correctly and *when* they are appropriate to use. This knowledge is crucial for daily life, as calculators are now a standard way of handling numerical operations, both at work and at home.

Using a calculator correctly is not a simple task; it depends on a good knowledge of the four operations and of the number system, so that students can select suitable calculations and also determine what a reasonable result would be. These skills are the basis of any work with numbers, whether or not a calculator is involved.

Unfortunately, calculators are often seen as tools to check computations with, as if other methods are somehow more fallible. Students need to understand that any computational method can be used to check any other; it's just as easy to make a mistake on the calculator as it is to make a mistake on paper or with mental arithmetic. Throughout this curriculum, we encourage students to solve computation problems in more than one way in order to double-check their accuracy. We present mental arithmetic, paper-and-pencil computation, and calculators as three possible approaches.

In this curriculum we also recognize that, despite their importance, calculators are not always appropriate in mathematics instruction. Like any tools, calculators are useful for some tasks, but not for others. You will need to make decisions about when to allow students access to calculators and when to ask that they solve problems without them, so that they can concentrate on other tools and skills. At times when calculators are or are not appropriate for a particular activity, we make specific recommendations. Help your students develop their own sense of which problems they can tackle with their own reasoning and which ones might be better solved with a combination of their own reasoning and the calculator.

Managing calculators in your classroom so that they are a tool, and not a distraction, requires some planning. When calculators are first introduced, students often want to use them for everything, even problems that can be solved quite simply by other methods. However, once the novelty wears off, students are just as interested in developing their own strategies, especially when these strategies are emphasized and valued in the classroom. Over time, students will come to recognize the ease and value of solving problems mentally, with paper and pencil, or with manipulatives, while also understanding the power of the calculator to facilitate work with larger numbers.

Experience shows that if calculators are available only occasionally, students become excited and distracted when they are permitted to use them. They focus on the tool rather than on the mathematics. In order to learn when calculators are appropriate and when they are not, students must have easy access to them and use them routinely in their work.

If you have a calculator for each student, and if you think your students can accept the responsibility, you might allow them to keep their calculators with the rest of their individual materials, at least for the first few weeks of school. Alternatively, you might store them in boxes on a shelf, number each calculator, and assign a corresponding number to each student. This system can give students a sense of ownership while also helping you keep track of the calculators.

Using Computers

Students can use computers to approach and visualize mathematical situations in new ways. The computer allows students to construct and manipulate geometric shapes, see objects move according to rules they specify, and turn, flip, and repeat a pattern.

This curriculum calls for computers in units where they are a particularly effective tool for learning mathematics content. One unit on 2-D geometry at each of the grades 3–5 includes a core of activities that rely on access to computers, either in the classroom or in a lab. Other units on geometry, measurement, data, and changes include computer activities, but can be taught without them. In these units, however, students' experience is greatly enhanced by computer use.

The following list outlines the recommended use of computers in this curriculum:

Grade 1
Unit: *Survey Questions and Secret Rules*
    (Collecting and Sorting Data)
Software: Tabletop, Jr.
Source: Broderbund

Unit: *Quilt Squares and Block Towns*
    (2-D and 3-D Geometry)
Software: *Shapes*
Source: provided with the unit

Grade 2
Unit: *Mathematical Thinking at Grade 2*
    (Introduction)
Software: *Shapes*
Source: provided with the unit

Unit: *Shapes, Halves, and Symmetry* (Geometry
    and Fractions)
Software: *Shapes*
Source: provided with the unit

Unit: *How Long? How Far?* (Measuring)
Software: *Geo-Logo*
Source: provided with the unit

Grade 3
Unit: *Flips, Turns, and Area* (2-D Geometry)
Software: *Tumbling Tetrominoes*
Source: provided with the unit

Unit: *Turtle Paths* (2-D Geometry)
Software: *Geo-Logo*
Source: provided with the unit

Grade 4
Unit: *Sunken Ships and Grid Patterns*
    (2-D Geometry)
Software: *Geo-Logo*
Source: provided with the unit

Grade 5
Unit: *Picturing Polygons* (2-D Geometry)
Software: *Geo-Logo*
Source: provided with the unit

Unit: *Patterns of Change* (Tables and Graphs)
Software: *Trips*
Source: provided with the unit

Unit: *Data: Kids, Cats, and Ads* (Statistics)
Software: Tabletop, Sr.
Source: Broderbund

The software provided with the *Investigations* units uses the power of the computer to help students explore mathematical ideas and relationships that cannot be explored in the same way with physical materials. With the *Shapes* (grades 1–2) and *Tumbling Tetrominoes* (grade 3) software, students explore symmetry, pattern, rotation and reflection, area, and characteristics of 2-D shapes. With the *Geo-Logo* software (grades 3–5), students investigate rotations and reflections, coordinate geometry, the properties of 2-D shapes, and angles. The *Trips* software (grade 5) is a mathematical exploration of motion in which students run experiments and interpret data presented in graphs and tables.

We suggest that students work in pairs on the computer; this not only maximizes computer resources but also encourages students to consult, monitor, and teach one another. Generally, more than two students at one computer find it difficult to share. Managing access to computers is an issue for every classroom. The curriculum gives you explicit support for setting up a system. The units are structured on the assumption that you have enough computers for half your students to work on the machines in pairs at one time. If you do not have access to that many computers, suggestions are made for structuring class time to use the unit with five to eight computers, or even with fewer than five.

Assessment plays a critical role in teaching and learning, and it is an integral part of the *Investigations* curriculum. For a teacher using these units, assessment is an ongoing process. You observe students' discussions and explanations of their strategies on a daily basis and examine their work as it evolves. While students are busy recording and representing their work, working on projects, sharing with partners, and playing mathematical games, you have many opportunities to observe their mathematical thinking. What you learn through observation guides your decisions about how to proceed. In any of the units, you will repeatedly consider questions like these:

■ Do students come up with their own strategies for solving problems, or do they expect others to tell them what to do? What do their strategies reveal about their mathematical understanding?

■ Do students understand that there are different strategies for solving problems? Do they articulate their strategies and try to understand other students' strategies?

■ How effectively do students use materials as tools to help with their mathematical work?

■ Do students have effective ideas for keeping track of and recording their work? Does keeping track of and recording their work seem difficult for them?

You will need to develop a comfortable and efficient system for recording and keeping track of your observations. Some teachers keep a clipboard handy and jot notes on a class list or on adhesive labels that are later transferred to student files. Others keep loose-leaf notebooks with a page for each student and make weekly notes about what they have observed in class.

Assessment Tools in the Unit

With the activities in each unit, you will find questions to guide your thinking while observing the students at work. You will also find two built-in assessment tools: Teacher Checkpoints and embedded Assessment activities.

Teacher Checkpoints The designated Teacher Checkpoints in each unit offer a time to "check in" with individual students, watch them at work, and ask questions that illuminate how they are thinking.

At first it may be hard to know what to look for, hard to know what kinds of questions to ask. Students may be reluctant to talk; they may not be accustomed to having the teacher ask them about their work, or they may not know how to explain their thinking. Two important ingredients of this process are asking students open-ended questions about their work and showing genuine interest in how they are approaching the task. When students see that you are interested in their thinking and are counting on them to come up with their own ways of solving problems, they may surprise you with the depth of their understanding.

Teacher Checkpoints also give you the chance to pause in the teaching sequence and reflect on how your class is doing overall. Think about whether you need to adjust your pacing: Are most students fluent with strategies for solving a particular kind of problem? Are they just starting to formulate good strategies? Or are they still struggling with how to start? Depending on what you see as the students work, you may want to spend more time on similar problems, change some of the problems to use smaller numbers, move quickly to more challenging material, modify subsequent activities for some students, work on particular ideas with a small group, or pair students who have good strategies with those who are having more difficulty.

Embedded Assessment Activities Assessment activities embedded in each unit will help you examine specific pieces of student work, figure out what it means, and provide feedback. From the students' point of view, these assessment activities are no different from any others. Each is a learning experience in and of itself, as well as an opportunity for you to gather evidence about students' mathematical understanding.

The embedded assessment activities sometimes involve writing and reflecting; at other times, a discussion or brief interaction between student and teacher; and in still other instances, the creation and explanation of a product. In most cases, the assessments require that students *show* what they did, *write* or *talk* about it, or do both. Having to explain how they worked through a problem helps students be more focused and clear in their mathematical thinking. It also helps them realize that doing mathematics is a process that may involve tentative starts, revising one's approach, taking different paths, and working through ideas.

Teachers often find the hardest part of assessment to be interpreting their students' work. We provide guidelines to help with that interpretation. If you have used a process approach to teaching writing, the assessment in *Investigations* will seem familiar. For many of the assessment activities, a Teacher Note provides examples of student work and a commentary on what it indicates about student thinking.

Documentation of Student Growth

To form an overall picture of mathematical progress, it is important to document each student's work in journals, notebooks, or portfolios. The choice is largely a matter of personal preference; some teachers have students keep a notebook or folder for each unit, while others prefer one mathematics notebook, or a portfolio of selected work for the entire year. The final activity in each *Investigations* unit, called Choosing Student Work to Save, helps you and the students select representative samples for a record of their work.

This kind of regular documentation helps you synthesize information about each student as a mathematical learner. From different pieces of evidence, you can put together the big picture. This synthesis will be invaluable in thinking about where to go next with a particular child, deciding where more work is needed, or explaining to parents (or other teachers) how a child is doing.

If you use portfolios, you need to collect a good balance of work, yet avoid being swamped with an overwhelming amount of paper. Following are some tips for effective portfolios:

- Collect a representative sample of work, including some pieces that students themselves select for inclusion in the portfolio. There should be just a few pieces for each unit, showing different kinds of work—some assignments that involve writing, as well as some that do not.

- If students do not date their work, do so yourself so that you can reconstruct the order in which pieces were done.

- Include your reflections on the work. When you are looking back over the whole year, such comments are reminders of what seemed especially interesting about a particular piece; they can also be helpful to other teachers and to parents. Older students should be encouraged to write their own reflections about their work.

Assessment Overview

There are two places to turn for a preview of the assessment opportunities in each *Investigations* unit. The Assessment Resources column in the unit Overview Chart (pp. I-13–I-17) identifies the Teacher Checkpoints and Assessment activities embedded in each investigation, guidelines for observing the students that appear within classroom activities, and any Teacher Notes and Dialogue Boxes that explain what to look for and what types of student responses you might expect to see in your classroom. Additionally, the section About the Assessment in This Unit (p. I-21) gives you a detailed list of questions for each investigation, keyed to the mathematical emphases, to help you observe student growth.

Depending on your situation, you may want to provide additional assessment opportunities. Most of the investigations lend themselves to more frequent assessment, simply by having students do more writing and recording while they are working.

Combining and Comparing

Content of This Unit Students solve problems that involve comparison—for example, finding the difference between two weights, between record-holding statistics and their own statistics, between handfuls of beans, and between the number of school days and non-school days in a year. They also do problems in which they determine how amounts of time, money, and measurement are combined. For example: What can I buy for $2.50? If several children's heights add up to 318 inches, what could the individual heights be? Emphasis throughout is placed on students developing their own addition and subtraction strategies that make sense to them. Students are encouraged to use estimation and multiple strategies to double-check their work. The unit also includes number games and problems that build students' fluency with using hundreds as important landmarks in combining and comparing.

Connections with Other Units If you are doing the full-year *Investigations* curriculum in the suggested sequence for grade 3, this is the seventh of ten units. The unit can be used successfully at either grade 3 or grade 4, depending on the previous experience and needs of your students.

Some of the activities in this unit extend the work of two other grade 3 units, *Mathematical Thinking at Grade 3* and *Landmarks in the Hundreds*. These prior units lay the groundwork for effective combining and comparing strategies. Investigations and mathematical ideas similar to those in this unit are also found in the grade 4 unit, *Money, Miles, and Large Numbers*. The fourth grade unit works with larger numbers and involves students in more complex addition and subtraction situations.

Investigations Curriculum ■ Suggested Grade 3 Sequence

Mathematical Thinking at Grade 3 (Introduction)

Things That Come in Groups (Multiplication and Division)

Flips, Turns, and Area (2-D Geometry)

From Paces to Feet (Measuring and Data)

Landmarks in the Hundreds (The Number System)

Up and Down the Number Line (Changes)

▶ *Combining and Comparing* (Addition and Subtraction)

Turtle Paths (2-D Geometry)

Fair Shares (Fractions)

Exploring Solids and Boxes (3-D Geometry)

Investigation 1 ▪ Comparisons with Record Numbers

Class Sessions	Activities	Pacing
Sessions 1 and 2 (p. 4) HOW MANY CHILDREN IN YOUR FAMILY?	Comparing Family Size Close to 100 Homework: Ages	minimum 1 hr
Session 3 (p. 12) MORE RECORD COMPARISONS	Loooking at the Animal Data Oldest Living Relative Extension: Comparing Our Class with Guinness	minimum 1 hr

Mathematical Emphasis

- Comparing two numbers and developing strategies for determining their difference

- Developing ways of getting close to 100 by combining numbers

- Using landmark numbers (multiples of 10 and 100) to compare two quantities

Assessment Resources

Directions for Close to 100 (Teacher Note, p. 8)

Using Landmark Numbers for Comparing (Teacher Note, p. 10)

How Many More Children? (Dialogue Box, p. 11)

Developing Strategies for Addition and Subtraction (Teacher Note, p. 16)

Using Concrete Materials for Addition and Subtraction (Teacher Note, p. 17)

Materials

Interlocking cubes

Family letter

Student Sheets 1–2

Teaching resource sheets

Investigation 2 ▪ How Much Heavier or Lighter?

Class Sessions	Activities	Pacing
Session 1 (p. 20) WEIGHING FRUITS AND VEGETABLES	Making Predictions Using the Pan Balance Home Sets of Numeral Cards Homework: Close to 100	minimum 1 hr
Session 2 (p. 24) COMPARING THE WEIGHTS	Reweighing and Finding the Difference Discussing the Changes Assessment: Close to 100	minimum 1 hr

◑ **Ten-Minute Math ▪ Exploring Data**

Mathematical Emphasis	**Assessment Resources**	**Materials**

Mathematical Emphasis

- Developing conjectures about and making comparisons of how things change over time

- Comparing weights with a pan balance

- Finding how far a number is from the next multiple of 10 or multiple of 100

Assessment Resources

Assessment: Close to 100 (p. 25)

Materials

Chart paper

Markers

Pan balances

Paper clips

Fruits and vegetables

Paper plates

Scissors

Envelopes or rubber bands

Student Sheets 3–4

Teaching resource sheets

Investigation 3 ▪ Adding with Money, Inches, and Time

Class Sessions	Activities	Pacing
Sessions 1 and 2 (p. 28) HEIGHTS AND COUPONS	Choice Time: Finding Numbers to Make a Total Discussion 1: Strategies for Addition Discussion 2: Finding Numbers to Make a Total Homework: Different Ways to Add Extension: Inches and Feet Extension: Measuring in Centimeters	minimum 2 hr
Session 3 (p. 35) PLANNING A PARTY	Planning a Party Teacher Checkpoint: Addition Strategies Homework: Choose a Problem Extension: Planning a Class Activity	minimum 1 hr

◐ **Ten-Minute Math** ▪ **Exploring Data**

Mathematical Emphasis

- Solving addition problems with multiple addends and keeping track of the steps

- Developing a repertoire of addition strategies that rely on students' number sense and understanding of number relationships

- Recognizing and using standard addition notation while using approaches based on sound number and operation sense

- Exploring number relationships and using important equivalencies in time, money, and linear measure

- Using estimation to make good approximations

Assessment Resources

Teacher Checkpoint: Addition Strategies (p. 36)

Three Powerful Addition Strategies (Teacher Note, p. 37)

Keeping Track of Addition and Subtraction (Teacher Note, p. 39)

Materials

Interlocking cubes

Calculators

Yardsticks

Play money

Colored pencils, markers, or crayons

Student Sheets 5–9

Teaching resource sheets

Investigation 4 ▪ Working with Hundreds

Class Sessions	Activities	Pacing
Session 1 (p. 42) HANDFULS OF BEANS	Comparing Handfuls of Beans Looking at the Handfuls Data Homework: Handfuls at Home Extension: Adding the Left Hand Data	minimum 1 hr
Session 2 (p. 48) MORE HANDFULS	Handfuls from Home Recording Comparisons with Standard Notation Assessment: Handfuls Problems Homework: Class Handfuls and My Handfuls Extension: Ordering Handfuls Extension: Giant Handfuls	minimum 1 hr
Sessions 3 and 4 (p. 53) HUNDREDS OF PAPER CLIPS	Quick Images: How Many Paper Clips? Combining, Then Subtracting Choice Time: Working with Hundreds Homework: Problem Strategies	minimum 2 hr

◑ **Ten-Minute Math** ▪ **Estimation and Number Sense**

Mathematical Emphasis

- Developing and communicating strategies for combining and comparing quantities in the hundreds and thousands

- Using standard addition and subtraction notation to record comparison problems

- Using multiples of 100 as landmarks for adding and subtracting

- Collecting, recording, and graphing data

- Making predictions about data, describing and interpreting data

Assessment Resources

Line Plot: A Quick Way to Show the Shape of the Data (Teacher Note, p. 47)

Assessment: Handfuls Problems (p. 50)

Adding and Subtracting Go Together (Teacher Note, p. 52)

Materials

Dried beans

Paper clips

Overhead projector

Calculators

Colored pencils, markers, or crayons

Envelopes

Paste or glue sticks

Student Sheets 10–14

Teaching resource sheets

Investigation 5 ▪ Calendar Comparisons

Class Sessions	Activities	Pacing
Session 1 (p. 60) HOW MUCH LONGER?	How Much Longer? How Many Days Until Your Birthday? Assessment: Counting the Days	minimum 1 hr
Sessions 2 and 3 (p. 64) SCHOOL DAYS	Days In and Out of School Making International Comparisons What's Your Opinion? Choosing Student Work to Save Homework: School Days Around the World Problems Extension: Time Travel Extension: Writing About Your Opinion Extension: Taking a Survey	minimum 2 hr

◔ **Ten-Minute Math** ▪ **Estimation and Number Sense**

Mathematical Emphasis

- Exploring the mathematical characteristics of the calendar and using them to solve problems

- Solving complex problems by breaking them into manageable parts

- Examining how parts and the whole are related in addition and subtraction

Assessment Resources

Assessment: Counting the Days
(p. 62)

How Do You Count the Days?
(Dialogue Box, p. 63)

Choosing Student Work to Save
(p. 68)

Counting School and Non-School Days
(Teacher Note, p. 70)

Materials

Calendars for current year

School calendars for current year

Wall calendar for current year

Calculators

Interlocking cubes

Student Sheets 15–16

Teaching resource sheets

Following are the basic materials needed for the activities in this unit. Many of the items can be purchased from the publisher, either individually or in the Teacher Resource Package and the Student Materials Kit for grade 3. Detailed information is available on the *Investigations* order form. To obtain this form, call toll-free 1-800-872-1100 and ask for a Dale Seymour customer service representative.

Snap™ Cubes (interlocking cubes) or counters

Pan balances (1 per 5–8 students)

Play money: a supply of bills and coins (optional)

Numeral Cards (manufactured, or use blackline masters at the back of this book to make your own sets)

Yardsticks or measuring tapes (5–10 per class)

The Guinness Book of Records, 1993, compiled by Peter Matthews (Bantam Books, 1993): 1 or 2 copies (optional)

Beans, dried; containers for the beans

Fruit and vegetables to cut up for weighing

Paper plates

Calendars for the current year (1 per pair of students)

School calendars for the current year (1 per pair of students)

Wall calendar for display (optional)

Calculators

Scissors

Drawing paper; colored pencils, markers, or crayons

Chart paper

Envelopes or rubber bands (to hold Numeral Card decks, Paper Clip Problems, and Related Problem Sets)

Paper clips in boxes of 100 (1 box per student)

Overhead projector and transparencies

Paste or glue sticks

The following materials are provided at the end of this unit as blackline masters. A Student Activity Booklet containing all student sheets and teacher resources needed for individual work is available.

Family Letter (p. 78)
Student Sheets 1–16 (p. 79)
Teaching Resources:
   100 Chart (p. 81)
   Coupons (p. 89)
   Close to 100 Scoring Variation: Counting Track (p. 91)
   Quick Image Paper Clip Boxes (transparency master) (p. 97)
   Paper Clip Problems (p. 98)
   Related Problem Sets (p. 99)
   School Days Around the World (p. 102)
   Numeral Cards (p. 103)
   How to Play Close to 100 (p. 106)
   300 Chart (p. 107)
Practice Pages (p. 109)

A Note on Measuring Tools For measuring their heights in Investigation 3, Sessions 1 and 2, students will be using inches. You may want to extend the activity with measuring in centimeters. Tape measures and yardsticks or metersticks are easiest for students to use.

Some schools may have a combination measuring tool—metersticks that are 100 centimeters long, but that are also calibrated in inches on the reverse. These are fine for use with metric measure; however, be careful if you plan to use them for measuring in inches. They look like yardsticks, but they are actually a little more than 39 inches long. They can be confusing to students (and adults) who use this tool, expecting to measure things in 3-foot lengths.

To reduce confusion, try covering the extra 3 inches with masking tape when students need yardsticks. When you distribute these tools, explain why you have covered the end. If you can get separate yardsticks and metersticks, use these instead of the combination stick.

Related Children's Literature

Kuskin, Karla. *The Philharmonic Gets Dressed.* New York: Harper and Row, 1982.

Levinson, Riki. *Country Dawn to Dusk.* New York: Dutton Children's Books, 1992.

Williams, Vera B. *A Chair for My Mother.* New York: Greenwillow Books, 1982.

This unit is about addition and subtraction in contexts in which two or more quantities are compared, one quantity is removed from another, or quantities are added together to reach a given total. The investigations are not based on straightforward, immediately obvious addition or subtraction situations. Rather, students need to think about what's actually involved in the problem:

> If my grandfather is 89 and the oldest person on record is 120, how much longer would my grandfather need to live to tie the record?

Does "how much longer" suggest that something should be added on? If so, what should be added to what? Can this problem be seen as either addition or subtraction? How would you solve it? How can it be represented using standard addition or subtraction notation? A key goal of this unit is for students to learn how to use addition and subtraction flexibly to solve problems.

Students' own strategies—strategies that involve good mental arithmetic and an understanding of number relationships—are what matter here. Students must develop combining and comparing strategies that make sense to them. For example:

> How would you compare an orange slice that weighs 52 paper clips with one that weighs 28 paper clips? How much more is 52 than 28?

A student might use a combination of addition and subtraction: "I know that 30 to 50 is 20. 28 is 2 below 30, and 52 is 2 above 50, so add those four onto the 20—it's 24. So it's heavier by 24 paper clips." This strategy is based on sound understanding of the relationship between 28, 52, and some key landmarks in the number system—in this case, multiples of 10.

One of the most important goals of the unit is to encourage students to articulate, develop, and use their own strategies for solving comparing and combining problems. As they develop and share these strategies with their peers, they are also encouraged to apply them to many different situations. It is important to see that the same strategy that worked for comparing weights of fruit will work for comparing ages in a family, or the number of days that children are in and out of school.

During this unit, students will be thinking a great deal about relationships between numbers or quantities without relying on memorized procedures. While there is nothing wrong with knowing the computational procedures that have traditionally been taught in American schools, this unit has a different emphasis. Too often, students memorize procedures that they don't understand, then apply them blindly without engaging in any mathematical thought. When the emphasis is on learning rote procedures, many students just try to follow the steps and seem to forget everything they know about relationships among numbers. They don't estimate, they don't think about the situation, and they don't check their results.

We want students to develop procedures that they can rely on because they are based on understandings about numbers and number relationships. For example, problems like the following are often particularly difficult for students who use the standard subtraction algorithm:

$$\begin{array}{r} 1001 \\ -3 \\ \hline \end{array}$$

The regrouping procedure that is typically taught in school is unnecessarily complex for this computation. In fact, for the child who understands how these numbers relate to each other and sees 1000 as a meaningful landmark in the number system, this problem is quite easy.

One of your most important tasks as a teacher during this unit will be to make sure that students only use approaches that they understand and can explain. You may need to insist that students show you how they are thinking about a problem by using concrete objects or the 300 chart. And it may take some time before they believe that you value their thinking, that you don't expect them to apply rote procedures, and that you want them to use what they know about number relationships to reason about addition and subtraction situations.

Mathematical Emphasis At the beginning of each investigation, the Mathematical Emphasis section tells you what is most important for students to learn about during that investigation. Many of these mathematical understandings and processes are difficult and complex. Students gradually learn more and more about each idea over many years of schooling. Individual students will begin and end the unit with different levels of knowledge and skill, but all students will develop strategies that make sense to them when adding and subtracting, will learn to apply these strategies to problem situations, and will recognize the standard notation that describes these situations. Students will also learn to keep track of their strategies by making notes and writing down intermediate steps, to use estimation as a check for the reasonableness of their results, and to double-check their work by using more than one approach to solve each problem.

Throughout the *Investigations* curriculum, there are many opportunities for ongoing daily assessment as you observe, listen to, and interact with students at work. In this unit, you will find one Teacher Checkpoint:

Investigation 3, Session 3:
Addition Strategies (p. 36)

This unit also has three embedded assessment activities:

Investigation 2, Session 2:
Close to 100 (p. 25)

Investigation 4, Session 2:
Handfuls Problems (p. 50)

Investigation 5, Session 1
Counting the Days (p. 62)

In addition, you can use almost any activity in this unit to assess your students' needs and strengths. Listed below are questions to help you focus your observation in each investigation. You may want to keep track of your observations for each student to help you plan your curriculum and monitor students' growth. Suggestions for documenting student growth can be found in the section About Assessment (p. I-10).

Investigation 1: Comparisons with Record Numbers

■ How do students find the differences between two numbers? Do they count up from the smallest number? use materials or representations? use landmark numbers? Do they use subtraction to find the difference?

■ What strategies do students use for Close to 100? Do they make use of the landmark numbers?

Investigation 2: How Much Heavier or Lighter?

■ How do students make conjectures about how things will change over time? Do they use evidence? How do they compare how different things change over time?

■ What strategies do students use to determine the difference between original and ending weights?

■ How do students find the distance a number is from the next multiple of 10 or 100? Are they developing fluency with this strategy while they play Close to 100?

Investigation 3: Adding with Money, Inches, and Time

■ What strategies do students use to solve addition problems with multiple addends? How do students keep track of their work?

■ What strategies are students developing for solving addition problems? Can they approach a problem in more than one way? Do they use one strategy to double-check another? Do their strategies reflect a developing number sense? an understanding of number relationships?

■ How do students recognize, interpret, and use standard notation? What strategies do students use to solve problems presented in standard notation? Do they use strategies that make sense?

■ When adding in the contexts of time, money, and linear measure, what strategies do students use? Do they use number relationships? important equivalencies?

■ How do students approximate a reasonable answer to a problem? How do they adjust their approximation to find an exact solution?

Investigation 4: Working with Hundreds

■ How do students read, write, combine, and compare numbers in the hundreds and thousands? Can they communicate those strategies orally? in writing?

■ How do students add and subtract large numbers? Are they using landmarks in the number system, such as multiples of 10 and 100?

■ How do students go about collecting and recording data? Do they have an organized or systematic approach? Do they ask everyone? How do they describe and represent their data?

Investigation 5: Calendar Comparisons

■ How do students count on the calendar? For example, do they rely solely on counting by 1's, or do they break time into months or weeks? Can they count backward as well as forward?

■ How do students keep track of their work, particularly in a complex problem with many parts? How do they double-check their work?

■ How do students relate the parts and the whole in addition and subtraction problems? Do they recognize that together the parts should equal the whole?

In the *Investigations* curriculum, mathematical vocabulary is introduced naturally during the activities. We don't ask students to learn definitions of new terms; rather, they come to understand such words as *factor* or *area* or *symmetry* by hearing them used frequently in discussion as they investigate new concepts. This approach is compatible with current theories of second-language acquisition, which emphasize the use of new vocabulary in meaningful contexts while students are actively involved with objects, pictures, and physical movement.

Listed below are some key words used in this unit that will not be new to most English speakers at this age level, but may be unfamiliar to students with limited English proficiency. You will want to spend additional time working on these words with your students who are learning English. If your students are working with a second-language teacher, you might enlist your colleague's aid in familiarizing students with these words, before and during this unit. In the classroom, look for opportunities for students to hear and use these words. Activities you can use to present the words are given in the appendix, Vocabulary Support for Second-Language Learners (p. 75).

record, ties   Students compare the number of children in their own family with the record number of children in *The Guinness Book of Records*. Then they determine how many more children there would need to be in their family to tie the Guinness record.

weight, balance   After learning how to use a pan balance, students weigh and reweigh pieces of fruits and vegetables over a period of time. They then determine the difference between the two weights for each kind of produce.

coupon   Students use grocery coupons to solve problems with multiple addends, in which a total amount of money is given and they must figure out a set of numbers that make that sum. For example: Find coupons that add up to $3.70.

hours, minutes   In planning a schedule for a two-hour party, students set starting times for various activities and decide how many minutes each activity will take.

birthday   Students use calendars to figure out how many days until their next birthday.

Multicultural Extensions for All Students

Whenever possible, encourage students to share words, objects, customs, or any aspects of daily life from their own cultures and backgrounds that are relevant to the activities in this unit. For example:

- When they are comparing the numbers of children in different families in Investigation 1, students may be interested in making more comparisons involving their own families. For example, they might make family trees, showing how many children were in the families of their parents and grandparents.

- Ask students about the kinds of fruit and vegetables they eat at home. Encourage them to bring in items that may not be familiar to other students, and use these as part of the weighing experiment in Investigation 2. In some areas, you might be able to visit stores that specialize in groceries for a particular culture represented by students in your classroom.

- Extend the international comparison of school days in Investigation 5 by researching information about the number of school days in the countries of origin of your students or their families. Students' relatives may be able to contribute information.

Investigations

Comparisons with Record Numbers

What Happens

Sessions 1 and 2: How Many Children in Your Family? Students compare their own data on the number of children in a family with the record number of children listed in *The Guinness Book of Records*. Students are also introduced to Close to 100, a game they will play throughout the unit, in which they compose a number that is close to 100 and compare this number with 100.

Session 3: More Record Comparisons Students bring in data from home and compare it with world record data. They compare the record ages of animals and people with the ages of their own pets and relatives.

Mathematical Emphasis

■ Comparing two numbers and developing strategies for figuring out the difference between them

■ Developing ways of getting close to 100 by combining numbers

■ Using landmark numbers (multiples of 10 and 100) to help in making a comparison of two quantities

What to Plan Ahead of Time

Materials

- Interlocking cubes or counters (along with the 100 charts and 300 charts you will be duplicating) should be available to all students throughout the unit for use in modeling problems.

- *The Guinness Book of Records, 1993* compiled by Peter Matthews (Bantam Books, 1993): 1 or 2 copies (Sessions 1–3, optional). A new edition is published every year and your students might have fun finding the latest issue, but the statistics we use here have not changed for many years.

- Numeral Cards: 1 deck per 2–3 students (Session 1). If you do not have manufactured sets, make your own; see Other Preparation.

Other Preparation

- Duplicate student sheets and teaching resources (located at the end of this unit) in the following quantities. If you have Student Activity Booklets, copy only the item marked with an asterisk.

For Sessions 1–2

100 chart (p. 81): 1 per student

Numeral Cards (p. 103): enough to make one deck per 2 or 3 students (see below)

How to Play Close to 100 (p. 106): 1 to post, or 1 per student, optional

Student Sheet 1, Close to 100 Score Sheet (p. 79): 1 per student

Student Sheet 2, Ages (p. 80): 1 per student (homework)

Family letter* (p. 78): 1 per student. Remember to sign it before copying.

For Session 3

300 chart (p. 107): 1 per student

- If you haven't purchased the *Investigations* grade 3 materials kit, make a deck of Numeral Cards for every two or three students. If you can duplicate the sheets on tagboard, the cards will last longer. Cut apart the 44 cards for one complete deck. If students help with cutting, many sets can be made quickly. Mark the back of each deck differently, or copy the sheets for each deck on paper of a different color, so that the decks are easy to separate if they get mixed up in use. (Sessions 1–2)

- Play the Close to 100 game by yourself or with someone else, so that you are familiar with the rules and scoring. (Sessions 1–2)

- If you plan to provide folders in which students will save their work for the entire unit, prepare these for distribution during Sessions 1 and 2.

- If you have pets in your classroom or school, find out how old they are. (Session 3)

- There is interesting background material about centenarians in *The Guinness Book of Records*, including information on countries with the highest rates of longevity. You may want to share this with students. (Session 3)

How Many Children in Your Family?

Materials

- Interlocking cubes
- 100 charts
- Numeral Cards (1 deck per 2–3 students)
- Student Sheet 1 (1 per student)
- How to Play Close to 100 (1 to post, or 1 per student, optional)
- Student Sheet 2 (1 per student, homework)
- Family letter (1 per student)
- *The Guinness Book of Records* (1 or 2 copies, optional)

What Happens

Students compare their own data on the number of children in a family with the record number of children listed in *The Guinness Book of Records*. Students are also introduced to Close to 100, a game they will play throughout the unit, in which they compose a number that is close to 100 and compare this number with 100. Student work focuses on:

- developing strategies for comparing two numbers
- exploring combinations of numbers that make 100
- using landmark numbers while making comparisons

Comparing Family Size

Explain to the class that they will be making many comparisons during this unit. One of the things they will do is make comparisons with *The Guinness Book of Records*:

Today you are going to compare the number of children in your family to the record number of children in a family. How should we count the number of children in our families? Whom should we count?

Students—particularly those with step-siblings and blended families—will probably have questions about which people to count. Defining how to count children in the family is an important mathematical task. Allow some time to discuss and decide who should be counted and help students arrive at a consensus. Questions like these may come up:

I'm a foster child. Which family should I count?

My mom's pregnant. Can I count the new baby?

My mom has a new husband and he has two kids that live with us. Do I count them?

My dad has a kid from his old wife. Should I count her?

My sister's married. Does she count?

These questions are not only legitimate, they are an essential first step in establishing a definition and a basis for comparison. Encourage children to talk about the questions and arrive at a consensus concerning the answers.

Even though students are often eager to share information, be sensitive to any who may feel uncomfortable with this discussion. It is important to respect different situations as you help students figure out how to count the number of children in their families.

Students may decide that they should use the same definition as *The Guinness Book of Records*, in order to make the comparison fair. If they decide this on their own, tell them that the *Guinness* definition includes all children who have the same birth mother.

Once students have decided on their definition of "children in a family," have them determine this number for their own family and write it on a sheet of paper. Ask for their estimates of the record number of children in a family, then reveal the following fact from *The Guinness Book of Records:*

The largest number of children in a single family was 69. This family lived in Russia about 250 years ago. It included 16 pairs of twins, 7 sets of triplets, and 4 sets of quadruplets—there were no single births.

How many more children would you need in your family in order to tie the record? In other words, what is the difference between the number of children in your family and the record number of children?

Have students work in pairs, comparing their individual family data with the record. They should have cubes and 100 charts to work with. As students work, circulate and ask them to explain how they are finding the difference. Encourage them to use materials and representations to show their strategies. See the **Teacher Note**, Using Concrete Materials for Addition and Subtraction (p. 17), for more on how to help students use cubes and charts effectively to solve problems.

Comparing the Largest Family with the Class When pairs of students have finished this comparison, add another one:

Imagine that our class is a family and you are all the children in the family. We have a large number of children here today, but not as many as in the record family. What's the difference between the number of children in this room today and the number of children in the record family?

This is a good time to start using the word *difference* in talking about comparisons. Students are often confused by this word because it has so many meanings in ordinary language. Ask students if they know what you mean when you say "the difference between the number of children in my own family and the record number of children."

Help students see that there are many meanings for this word, but that there is a particular mathematical meaning—the *difference* between two quantities asks how much more there is of one thing than another; for example, the difference between my height and your height, the difference between the number of people in my family and your family, the difference between the number of red cars and blue cars in the parking lot. As you continue to use the word *difference* naturally throughout this unit, students will come to understand its mathematical use.

Sharing Findings   Ask several volunteer pairs to share their findings with the class, explaining how they worked out their comparisons. Encourage students to share the different strategies they used. This exchange is important in establishing the fact that there are different ways of comparing two numbers. See the **Dialogue Box**, How Many More Children? (p. 11), for an example of such a discussion.

Some students may see this comparison as a subtraction problem, while others may not. Do not push the use of subtraction, as adding on from a smaller number to a larger one is a perfectly reasonable approach to these problems and may make more sense to many of your students. See the **Teacher Note**, Adding and Subtracting Go Together (p. 52). Be sure to ask students who have used subtraction to say why they think it works.

Tell students that they will be gathering some data at home to compare with the record ages for the oldest pet and the oldest person in the next session.

Activity

Close to 100

Explain how to play the game Close to 100. The **Teacher Note**, Directions for Close to 100 (p. 8), gives step-by-step directions and examples of a few rounds of play. If you like, post or hand out to students the sheet, How to Play Close to 100 (p. 106).

This game focuses on finding pairs of two-digit numbers which make a total that is as close to 100 as possible. Knowing the difference between a number and the next landmark in the number system, such as a multiple of 10 or 100, is particularly useful in addition and subtraction. The **Teacher Note**, Using Landmark Numbers for Comparing (p. 10), explains how the type of thinking students do in Close to 100 will help them develop good mental arithmetic strategies.

After you have introduced the game, students play together in pairs or groups of three. Each group will play with one deck of Numeral Cards, and students will each need at least one copy of Student Sheet 1, Close to 100 Score Sheet, to keep track of their scores for each game (each sheet has space for two games). At this time, students should play only the basic game; the scoring variation (described in the Teacher Note) will be introduced later in the unit.

Note: Be sure to save the decks of Numeral Cards for use later in the unit.

Sessions 1 and 2 Follow-Up

Ages Send home copies of the family letter or the *Investigations* at Home booklet and Student Sheet 2, Ages. Students will collect their own data and determine the difference between the ages of animals they know about and the record ages listed on the sheet. If students don't have pets, encourage them to interview neighbors or friends who do to collect some names and ages. Those who live in more rural areas can also gather information about farm animals.

 **Homework**

❖ **Tip for the Linguistically Diverse Classroom** Students can draw any pets they want to list on Student Sheet 2.

This student sheet also asks students to find out the age of the oldest person in their family, which they will compare with the record age during Session 3. Explain that the oldest person could be any living relative—a grandparent, great-grandparent, uncle, aunt, great-uncle or aunt, or anyone else in the family. Remind them that they'll need the data on this sheet for Session 3.

As students come to school the next day, have them list their pet data on the board or on chart paper so that you will have it consolidated in a table for a class discussion (see example on p. 12).

Close to 100 can be played as a solitaire game, but in class, two or three students will play together. Each group will need one deck of Numeral Cards and a Close to 100 Score Sheet (Student Sheet 1) for each player.

How to Play

1. For the first round, deal out six cards to each player.

2. Each player uses any four of these cards to make two numbers that, when added, come as close as possible to a total of 100. (See the sample round below.) Wild Cards can be used for any numeral.

3. The player records these two numbers and the total on Student Sheet 1. The player's score for each round is the difference of the sum of the two numbers from 100. The four cards used are then placed in a discard pile.

4. For each successive round, four new cards are dealt to each player, so that all players again have six cards.

The game ends after five rounds. If the deck runs out of cards before the game is over, shuffle the discard pile and continue to deal. At the end of five rounds, players total their scores. The *lowest* score wins.

Sample Game

Round 1

Cesar is dealt these cards:

| 5 | 8 | 6̲ | 9̲ | 2 | 7 |

Annie is dealt these cards:

| 9̲ | 1 | 5 | 5 | 4 | 7 |

Cesar makes 58 + 29, and Annie makes 45 + 57.

Round 2

Cesar has 6 and 7 left from round 1, and is dealt

| 3 | 6̲ | 9̲ | 2 |

Annie has 9 and 1 left from round 1, and is dealt

| 8 | 2 | 5 | 0 |

Cesar makes 36 + 62, and Annie makes 98 + 02.

Note: Both Cesar and Annie could have gotten closer to 100 in round 1, and Cesar could have gotten closer to 100 in round 2. Can you see how?

The game proceeds, and their final scores look like this:

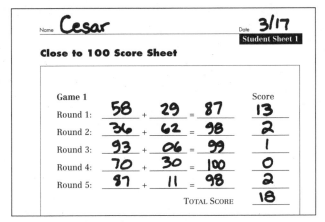

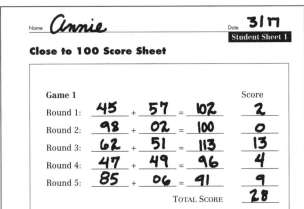

Cesar has the lowest score, so he wins.

Scoring Variation: Negative and Positive Integers

Students should be very comfortable with the basic game before trying this variation. Its use is specifically suggested during Choice Time in Investigation 3.

In this variation, the game is scored with negative and positive integers. If a player's total is above 100, the score is recorded as positive. If the total is below 100, the score is negative. For example, a total of 103 is scored as +3 (3 above 100) while a total of 98 is scored as –2 (2 below 100). If using this variation, Cesar's and Annie's score sheets from the sample game would look like this:

The player with the total score closest to zero wins. So, in this case, Annie wins (+2 is 2 away from 0, and –18 is 18 away from zero). Scoring this way changes the strategy for the game. Even though Cesar got many scores very close to 100, he did not compensate for his negative values with some positive ones. Annie had totals farther away from 100, but she balanced off her negative and positive scores more evenly to come out with a total score closer to zero.

The Close to 100 Scoring Variation: Counting Track (p. 91) can help players figure their total scores at the end of five rounds. For example, Annie's scores are +2, 0, +13, –4, and –9. To add these, she would start with a counter on 0 on the counting track, then move forward for positive scores and backward for negative scores—counting forward 2, no change, forward 13, back 4, and back 9. The total score is the last number the counter lands on—in Annie's case, +2.

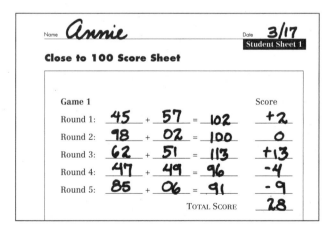

CLOSE TO 100 SCORING VARIATION: COUNTING TRACK

–23	–24	–25	–26	–27	–28	–29	–30	–31	–32
–22									
–21	–20	–19	–18	–17	–16	–15	–14	–13	–12
									–11
–1	–2	–3	–4	–5	–6	–7	–8	–9	–10
0									
+1	+2	+3	+4	+5	+6	+7	+8	+9	+10
									+11
+21	+20	+19	+18	+17	+16	+15	+14	+13	+12
+22									
+23	+24	+25	+26	+27	+28	+29	+30	+31	+32

91

*Investigation 3 • Resource
Combining and Comparing*

Number sense involves a deep understanding of numbers, their characteristics, their place in the number system, and their relationships to each other. Think for a minute about this problem:

> If you are 48 years old and I am 62, how much older am I than you?

When you think about finding the difference between 48 and 62, you immediately bring your number sense into play. You recognize very quickly a great deal about these numbers. You might use any of a number of ideas—that 62 is 2 more than 60, that 48 is 2 less than 50, that 48 + 10 is 58, that 62 − 10 is 52, that 10 is the difference between 50 and 60.

A lot of the information we use to solve this kind of problem mentally has to do with the relationship between a quantity in the problem and a nearby landmark in the number system. For example, we could solve this problem by thinking that 48 to 50 is 2, 50 to 60 is 10, 60 to 62 is 2 more, so the difference is 10 + 2 + 2 or 14. We used multiples of ten—50 and 60—as critical landmarks or anchors to which the numbers 48 and 62 are connected.

One purpose of this unit is for students to become more fluent with the use of such landmarks in the number system to solve addition and subtraction problems. As their work moves into numbers in the hundreds, multiples of 100 will join multiples of 10 as critical landmarks. In Investigation 5, for example, students will compare the number of school days in Hong Kong (220) with the number of school days in the United States (180). In this case, the number 200 can be used as an important landmark: from 180 to 200 is 20, and from 200 to 220 is 20, so the difference between 180 and 220 is 20 + 20 or 40.

In order to use these critical numbers effectively, students must become fluent in finding the difference between a number they are using and nearby landmarks. The game Close to 100 provides practice in this skill. (If I have 48, how much more do I need to make 100? If I have made a sum of 87, how far away is that from 100?) As students play this game, encourage them to use what they know about number relationships. If a student is trying to figure out what is needed to add to 48 to get 100, he or she might count by 10's: "58, 68, 78, 88, 98—that's five 10's, and 2 more is 52." Another student might use 50 as a landmark: "I know 50 to 100 is 50, and 48 to 50 is 2, so it's 52."

Powerful mental arithmetic strategies are based on this kind of knowledge. As students become more fluent in using what they know about number relationships and critical landmarks in the number system, they will develop strategies they can rely on to solve addition and subtraction problems.

How Many More Children?

In this discussion during the activity Comparing Family Size (p. 4), students share strategies for finding the difference between the 69 children in the "world's largest family" and the number of children in their families and in their class.

Seung, how did you solve the problem about your own family?

Seung: My family has 3 children, and I wanted to get up to 69. So I knew I needed 6 more to get to 9, then I needed 60 more to get to 69—6 and 60 is 66, so it's 66.

Sounds like you were thinking about making the numbers easier by using tens. Who has other ideas about how to find the difference between 69 and 3?

Ryan: We have 3, too. I'd just say 69 – 9 is 60. That would be nine. Then I'd go down from 60 to 10. That's 50 more, so it's 59. Then I'd go down 7 more to 3. And 59 + 7 is 66.

Latisha: That's sort of the same way I did it, only backwards.

The difference works out the same, going backward or forward. What about working with some other numbers—who has a big family with lots of children?

Kate: We have 7. And I knew if we had 9 children, we'd have 60 more to go to tie the record. 7 is 2 away from 9. So it's 62 more children to go for us. But I don't think my mom wants any more kids.

What about comparing the number of children in class today with the number of children in the record family? How did you find the difference between these numbers?

Ly Dinh: I counted up from 24 to 69 on the 100 chart.

How did you count?

Ly Dinh: By 1's, like 25, 26, 27, 28, and I kept going and it was 45.

Sometimes counting by 1's can be really hard because it's easy to lose track. Did anyone have a different way to count on the 100 chart?

Chantelle: I went like this: 34, 44, 54, 64— because every time you jump up a line it's 10, so that was 40, and then I just counted 5 more to 69.

Samir: I got 45, but I didn't get it that way. I counted up how many between 30 and 70, which is 40. Then I took 1 away, because the number's really 69, not 70. That's 39. Then I added on 6, because going from 24 to 30 is 6.

More Record Comparisons

Materials

- Cubes, 100 charts, 300 charts
- Completed Student Sheet 2 (from homework)
- *The Guinness Book of Records* (optional)

What Happens

Students bring in data from home and compare it with world record data. They compare the record ages of animals and people with the ages of their own pets and relatives. Student work focuses on:

- developing strategies for comparing two numbers
- using landmark numbers while making comparisons

Activity

Looking at the Animal Data

Earlier in the day, students should have listed in a table, on the board or chart paper, the pet data they collected on Student Sheet 2. You may also want to add data about any school pets or your own pets. Help them organize the names and ages of the animals into columns:

Cats – age	Dogs – age	Gerbils – age	Mice – age	Fish – age
Mimi 10	Riva 4	Sam 1	spot 1	Goldie 1
misty 1	Tasha 1		Piggy 2	
Gus 4	Woody 12			
	Sprig 6			
	Barney 5			

At the beginning of class, direct attention to the completed table and ask:

What do you notice about the ages of these pets? Which pets seem to live the longest? Which ones don't live as long?

Students will probably also want to talk about factors contributing to longevity, such as diet, being fit, or not getting into fights. Delve into the scientific aspects of this discussion as time and interest allow.

Comparing Pet Data Ask students to examine the information about their animals and the ones that hold records (listed on Student Sheet 2) to determine whether anyone in the class has broken any *Guinness* records (probably not, but ask anyway).

For homework, you've all found the differences between the ages of the pets on your list and the oldest ones on record. For each pet that *Guinness* has records for, let's figure out which one in this class comes closest to tying the record.

Here are the records from the 1993 *Guinness Book of Records* listed on the student sheet:

Guinea pig	14 years
Dog	29 years
Cat	36 years
Gerbil	8 years
Mouse	7 years
Rabbit	18 years

Note that *Guinness* doesn't provide records for all pets. Students may want to discuss why this is so. A fish, for example, is a common pet, but given how hard it is to identify individual fish, it would be very hard to get accurate information on the record age.

As you compare the *Guinness* data and the students' data, focus the discussion on the following:

- What two numbers are being compared?
- How many more years would the oldest pet in the class have to live to tie the world record?
- What are students' strategies for finding the difference?

Oldest Living Relative

Students now go on to compare the oldest person who ever lived with their own oldest relatives. Some will ask if they can list a relative who is dead, but lived to be very old. The class can decide whether this is acceptable.

Ask students how old they think the oldest person was when he or she died. Then explain:

According to the 1993 *Guinness*, the oldest person in the world died when he was 120 years old. The man was born in Japan more than a

**century ago. He died of pneumonia in 1986. Here is the comparison that
we will be making: How many more years would your oldest relative
have to live in order to tie this record?**

Using their data from Student Sheet 2, students work on these comparisons
in pairs, then share and discuss them with the class. A supply of cubes, 100
charts, and 300 charts should be available for them to work with. If some
students do not have their own data, they can pair up with those who do.
Then, as a class, figure out who has the oldest relative, and how much
longer this person would have to live to tie the record of 120 years.

In this comparison, students are likely to be comparing a number less than
100 with the number 120. In doing this, encourage them to use 100 as a
landmark to make the comparison. (See the **Teacher Note,** Using
Landmark Numbers for Comparing, p. 10.) By now, some students will have
picked up the idea that they can use subtraction to make comparisons and
find the difference between two numbers. This is a very important discov-
ery! You will want to encourage them to use subtraction, or combinations
of addition and subtraction, as appropriate. However, be wary of using the
subtraction algorithm in isolation, without good explanations of why and
how it works. The **Teacher Note,** Developing Strategies for Addition and
Subtraction (p. 16), explains why this is important.

Session 3 Follow-Up

Comparing Our Class with *Guinness*   This activity involves students in making comparisons between record holders in a recent copy of *The Guinness Book of Records* and their own class. Their own records and comparisons with the official ones will become a Class Record Book.

 **Extension**

Prepare for the activity by choosing several appropriate categories from the *Guinness Book* that students can easily investigate in class or for homework. Some possible choices are heaviest cat, oldest human, largest pumpkin (or tomato or peach), highest domino stack, and other categories like these. For each category, figure out who the record holder in the class is by having students bring information about their pets, or whatever category they select, to class. When the record is established, ask the owner to make a page for the Class Book. The page might include the category, a drawn picture or photograph of the record object, and the appropriate measurement (for example, 130 pounds of pumpkin).

After finding the class's record objects, students look in *The Guinness Book of Records* for the corresponding record. They add this to the page they have created for the Class Book, then figure out and write down the difference between the class record and the official record. Students may begin to predict what the relationship will be between records and may choose additional categories to investigate. The comparisons may be stated in terms of differences between the two records: For example, the largest strawberry in the world is 2 pounds heavier than the largest one we could find. The comparison may also be stated in terms of ratios between the two: For example, the largest tomato in the world is five times as large as the largest one in our class.

Developing Strategies for Addition and Subtraction

By third grade, students should be developing several different procedures that they can use fluently and flexibly to solve addition and subtraction problems. Some students will come to your class with previous experience developing their own procedures based on what they know about operations and number relationships. Others may know only the historically taught column addition and subtraction, using carrying and borrowing. Students who have learned these traditional algorithms by memorizing them and who are not used to thinking through the structure of story problems may have a difficult time at first with some of the problems in this unit.

Many real mathematical problems do not shout out "I am *this* kind of problem! Use *this* operation to solve me!" In fact, when many students first encounter some of the comparison problems in this unit, they say, "Just tell me whether I should add or subtract." Students have learned that when they have two numbers, they need to do an operation to get the right result, but they may not be used to thinking about why and how addition or subtraction can be applied to find out, for example, how much older a 94-year-old is than a 79-year-old. Students who are not used to thinking through the structure of a problem may need help in visualizing the relationships in the problem. Encourage students to tell you everything they know about the problem, to draw pictures to show what they know and what they need to find out, or to close their eyes and visualize what is happening in the problem.

Once students can describe the structure of a problem, they are ready to choose an operation to solve the problem. Support students to develop their own strategies based on what they know about the numbers and operations involved in the problem. Encourage students to make an estimate first, to think about the tens and ones in the problem, and to notice what landmark numbers (such as multiples of 10 or 25) the numbers in the problem are close to. Suppose the problem involves adding up three 17's (17 + 17 + 17). Many competent math

users—children as well as adults—solve this problem by adding the tens first:

> 10, 20, 30, then two 7's is 14, so 30 + 10 + 4 is 44, then 7 more is . . . let's see, 6 more would be 50, so it's 51.

Or they might use nearby landmarks:

> Well, 17 is close to 20, three 20's is 60, then subtract 3 for each 17, so minus 9 is 51.

Of course, students may still make errors and must still check for accuracy, but if they use methods that make sense to them, they are more likely to be able to find their own mistakes.

It's fine for students to use the traditional addition algorithm using "carrying" if they understand clearly how and why they are using it. Students who are simply using this procedure without thinking about the numbers in the problem may make the following error:

$$\begin{array}{r} {}^{1}17 \\ 17 \\ + 17 \\ \hline 42 \end{array}$$

These students are saying to themselves, "Put down the 2 and carry the 1." In their focus on the mechanics of this rule, they fail to see that they should instead "put down the 1 and carry the 2." To see if students understand how their addition and subtraction procedures model the situation, ask questions like these:

- Can you show me how that would work with the cubes or on the 100 chart?

- How would you show a first grader who doesn't understand what you just wrote how you solved this problem?

- Did you make an estimate? Is your answer close to your estimate? What other method can you use to double-check your solution?

For information on a variety of approaches that students developed based on sound number and operation sense, see the **Teacher Notes**, Using Landmark Numbers for Comparing (p. 10) and Three Powerful Addition Strategies (p. 37).

Using Concrete Materials for Addition and Subtraction

Many of your students will need to model problems in some way that helps them see more clearly the problem situation and what they are trying to find out. Help students use cubes, counters, 100 charts, or 300 charts to show you how they are thinking about the problem. The following examples show how students might use these materials:

Cubes or Counters   Suppose a student is comparing a family of 4 children to the 69 children in the "record" family. This student might build 69 with cubes—making 6 rows of ten and 1 row of nine, then put 4 connected cubes on top of the 69, then count how many more children there are in the group of 69.

Another student might see this problem as finding the difference between 69 and 4, and so might build 69 with the cubes, then take 4 away, and count how many are left. A third student might just imagine the 60, build 9, take away 4 cubes from the 9, and then mentally add back the 60.

As students are counting with cubes, gradually help them move beyond counting by 1's. Some students cling to counting by 1's because they feel this is the only way they can be sure of their accuracy. However, counting by 1's is difficult, inefficient, and prone to error—for anyone. Help students build confidence and skill in counting by 2's, 5's, and 10's. If you keep your interlocking cubes stored in towers of 10, students will be able to work with them more easily.

100 Chart and 300 Chart   Using the 100 chart to make the same comparison (the difference between 4 children and 69 children), students could place a counter or mark on 4 on the chart, and then find how many steps or jumps it will take to reach 69. At first, students may count by 1's. Encourage them to figure out how to make bigger jumps, such as from 4 to 10, then to 20, 30, 40, 50, 60, and finally to 69. Demonstrate how to jot down intermediate differences while jumping along the chart (6, 10, 10, 10, 10, 10, 9).

One of the biggest problems students have on the 100 chart is understanding where to start counting. When finding the difference between 69 and 4, for example, many students count both the 4 and the 69, ending up with a difference of 66 instead of 65. Help these students by working with much smaller numbers:

How many more children would you need in your family to make 5? 6? 7? Why don't you say 5 is two bigger than 4?

Students who persist with counting the beginning number should return to using cubes or counters. Reformulate the question in this way:

Here are the 4 children in your family. I'll make them with red cubes. I want you to add on blue cubes until you have 69. Then we can see how many extra children you'd need to have 69 in your family.

If some students are finding differences correctly on the 100 chart but are counting by 1's, help them begin to take bigger jumps by asking questions like these:

Where would you land if you took 10 steps? Now what if you took 10 more steps?

As students work with problems with larger numbers, they will find that the 300 chart can replace the 100 chart to help them visualize the relationships among the numbers.

How Much Heavier or Lighter?

What Happens

Session 1: Weighing Fruits and Vegetables
Students make predictions about what will happen to the weight of cut-up produce when it is exposed to air. They learn how to use a pan balance, then weigh and record the weights of various pieces of produce in terms of something with constant weight, such as paper clips. Students also make individual sets of Numeral Cards so that they can play Close to 100 at home.

Session 2: Comparing the Weights Students reweigh the produce one day or more after the initial weighing and determine the difference between the two weights for each kind of produce. Then they discuss what happened to the weight of the produce, why it happened, and how it varied with different types of produce. Students also work on an assessment of their knowledge about combinations of numbers that make 100.

Mathematical Emphasis

- Developing conjectures about and making comparisons of how things change over time
- Comparing weights with a pan balance
- Finding how far a number is from the next multiple of 10 or multiple of 100

Scheduling Notes

- Try to do this investigation when a weekend or holiday separates the two class sessions. While you can do these sessions on two consecutive days, leaving the produce for a longer time will mean a greater loss of weight and more dramatic data.

- If you don't have enough pan balances for all groups to work simultaneously, plan to have students take turns doing their weighing some other time during the same day. Each group will need approximately 15 minutes to weigh their food, and they need to do this for each of the two sessions in order to make comparisons.

What to Plan Ahead of Time

Materials

- Chart paper and markers for recording predictions (Session 1)
- Pan balances: 1 per group of 5–8 students (Sessions 1–2)
- Paper clips: 1–2 boxes for each pan balance (Sessions 1–2). You can substitute other small objects with constant weight; however, we have found that paper clips are sensitive enough to reflect the weight loss and are easily managed by students.
- A variety of ripe fruits and vegetables (Session 1, see below)
- Paper plates: 1 per group of 3–4 students (Session 1)
- Scissors (Session 1)
- Envelopes or rubber bands (for card decks): 1 per student (Session 1)

Other Preparation

- Duplicate student sheets and teacher resources (located at the end of this unit) in the following quantities. If you have Student Activity Booklets, no copying is needed.

For Session 1

Student Sheet 3, Weights and Comparisons (p. 82): 1 per student

Numeral Cards, pages 1–3 (p. 103): 1 set per student

How to Play Close to 100 (p. 106): 1 per student (homework)

Student Sheet 1, Close to 100 Score Sheet (p. 79): 2–3 per student (homework)

For Session 2

Student Sheet 4, Who Is Closer to 100? (p. 83): 1 per student

- Calibrate the pan balance you use to demonstrate, and show students how to do this. A simple way is to put 5 paper clips in each pan to see if the scale arm is level. If it's not level, adjust it. (Sessions 1–2)
- For Session 1, bring in (or have students bring in) produce that is past its prime and nearly ready to throw out. Choose produce with a high water content, such as lettuce, bananas, oranges, apples, apricots, pears, cucumbers, potatoes, and onions (feel free to add or substitute). Each group of students will need one small piece of each different kind. For a class of 24, you might use this amount:

2 bananas	2 apples	1–2 cucumbers
1 potato	1 onion	2 oranges
2 tomatoes	2 carrots	lettuce leaves

Just before students will weigh for the first time, cut the produce into small (less than bite-sized) pieces, making a piece of each item for each group. **Note:** Some produce loses weight even over a couple of hours, so it is very important to cut the pieces immediately before weighing them.

To avoid waste, use the leftovers to feed any guinea pigs, hamsters, or rabbits who inhabit the school. In the spirit of conservation, one child recommended that we use the remains of the project as food for fruit flies! The project remains would also be a nice addition to the compost heap.

Weighing Fruits and Vegetables

Materials

- Pan balances (1 per small group)
- Paper clips (1–2 boxes per pan balance)
- Cut-up fruits and vegetables
- Paper plates (1 per small group)
- Chart paper (optional)
- Student Sheet 3 (1 per student)
- Numeral Cards, pages 1–3 (1 set per student)
- Scissors
- Envelopes or rubber bands (1 per student)
- Student Sheet 1 (2–3 per student, homework)
- How to Play Close to 100 (1 per student, homework)

What Happens

Students make predictions about what will happen to the weight of cut-up produce when it is exposed to air. They learn how to use a pan balance, then weigh and record the weights of various pieces of produce in terms of something with constant weight, such as paper clips. Students also make individual sets of Numeral Cards so that they can play Close to 100 at home. Their work focuses on:

- developing conjectures about what will happen to the produce
- learning to weigh objects accurately, using a pan balance

Note: Some of the actual weighing can be done outside of math class if there are not enough balances to do this during class.

 Ten-Minute Math: Exploring Data Once or twice during the next few days, during ten minutes outside of math class, try this activity to give students continued opportunities to collect, graph, and describe real data. If you are using the year-long *Investigations* curriculum, you will find that this activity builds on ideas presented in previous units.

Choose a question about data that students know or can readily observe in class: How many buttons are you wearing today? What month is your birthday? Or, what is the main color in your clothes today?

As students report, quickly graph their data on a line plot or bar graph. (See the **Teacher Note**, Line Plot: A Quick Way to Show the Shape of the Data, p. 47.)

Ask students to describe the data: What do they notice? Where do most of the data seem to fall? What seems typical or usual for this class?

Ask them to interpret and make predictions: Why do you think that the data came out this way? Does anything about the data surprise you? Do you think we'd get similar data if we collected it tomorrow? next week? in another class? with adults? List any new questions that arise from considering these data.

For full instructions and variations, see p. 71.

Making Predictions

Begin this session by explaining that today the class will take a look at what happens to fruits and vegetables when they are left out for a while. Ask students to discuss their experiences with this:

What happens if you leave a half-eaten apple or orange in the car?

Encourage them to base their predictions on their own experiences with forgotten food, food that stays in the refrigerator too long, or food that's left in the sun. Students will probably comment on the physical appearance of food, as well as what happens to size and weight. Show them the pieces of fruits and vegetables and explain that they will weigh these pieces when they are fresh, and weigh them again another time to see what happens to their weight when they are left out for a while.

What do you think will happen to the *weight* of the food pieces if we leave them out for a day or more?

Elicit theories about the direction and amount of weight change, as well as why students think the changes might be happening.

Don't expect them to know that loss of water is the factor that's responsible for loss of weight, but do expect them to make predictions about what will happen, to observe what happens, and to theorize about why it might be happening. Students might think that food gets heavier over time—after all, when it's growing it gets heavier, and people and animals often get heavier over time. The prediction discussion will no doubt be lively. Focus students' attention on thinking about how and why the changes are occurring.

Write all predictions on the board or on chart paper and save them for Session 2.

Using the Pan Balance

Demonstrate how to use the pan balance. If students have not used one before, they may need a bit of practice before they do the actual weighing. Using paper clips and another object such as a pencil, show how the pan balance works:

Anything that you put in one side can be balanced with something in the other side. We will figure out how much something weighs by using paper clips to balance it. You can think about the weight of each object as how many paper clips it balances.

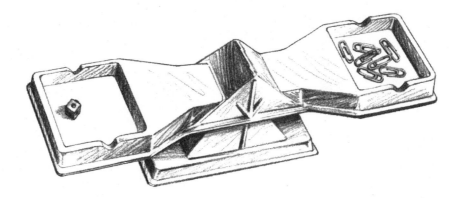

Have a student come up and weigh a few common items, using the paper clips to balance the weight. Ask questions like the following as this is done:

Watch the balance arm. How can you tell if it's balanced?

If the paper clip side goes down farther than the object, which side weighs more? What can you do to make the sides weigh the same?

Before weighing the pieces of food, ask students where they want to keep their produce between the first weighing and the second. Be sure that they put it in a relatively pest-free location. Ask questions to prompt their thinking about this:

Will the place you keep it make a difference in terms of how much weight it loses or gains? What will happen if you place the produce in a sunny spot? in the closet? in a refrigerator?

Students will work on the weighing in small groups of three or four. Distribute one copy of Student Sheet 3, Weights and Comparisons, to each student, and a paper plate to each group. Explain that each group will get a small piece of each kind of produce to weigh. They should weigh each piece of produce and record the results in the first two columns of their student sheets.

If you have enough pan balances, devote the rest of this session to the weighing. If you do not have enough, students who are not weighing can make their home sets of Numeral Cards (see below) while others are weighing. If necessary, have students continue to take turns weighing their produce after math class, so that all students can record their data on the same day.

When groups are finished weighing, they write their names on a paper plate, collect their pieces of produce on the plate, then set it in the chosen spot.

Home Sets of Numeral Cards

While some students are weighing, the rest of the class can be making individual decks of Numeral Cards for playing Close to 100 at home. Each student needs one copy of the Numeral Cards, pages 1–3. Students cut out the cards on these sheets to make one complete deck of 44 cards. Provide each student with an envelope or rubber band to keep the cards together. If students have scissors at home, you might have them do the cutting as part of their homework. Students can also play a game of Close to 100, if there is time.

Session 1 Follow-Up

Note: If you do not have enough pan balances for each group to reweigh their produce during the next math class, make arrangements for them to reweigh after at least a day has passed, but before proceeding with the activities in Session 2.

Close to 100   Students take home a deck of Numeral Cards (or copies of Numeral Cards, pages 1–3, to make a deck at home). They also take home two or three copies of Student Sheet 1, Close to 100 Score Sheet, and a copy of How to Play Close to 100. Ask students to teach the game to friends or family members and play it with them.

 **Homework**

Comparing the Weights

What Happens

Students reweigh the produce one day or more after the initial weighing and determine the difference between the two weights for each piece of produce. Then they discuss what happened to the weight of the produce, why it happened, and how it varied with different types of produce. Students also work on an assessment of their knowledge about combinations of numbers that make 100. Their work focuses on:

■ determining the difference between original and ending weights
■ developing theories to explain what happened to the weights
■ finding combinations that make 100

Materials

■ Pan balances
■ Paper clips
■ Plates of fruit and vegetable pieces (from Session 1)
■ Students' copies of Student Sheet 3 (partly completed in Session 1)
■ List of predictions (from Session 1)
■ Student Sheet 4 (1 per student)

Activity

Reweighing and Finding the Difference

At least one day after they have completed the initial weigh-in (at the beginning of this session or before), students weigh their produce a second time. Working in the same groups, they record the results of this second weighing on Student Sheet 3. Each group will need about 15–20 minutes for weighing. Students then calculate how much weight was lost or gained and write the difference on Student Sheet 3.

As students work, circulate and ask them about how they are determining the differences. Help students articulate their strategies. If they are automatically applying the standard subtraction procedure for each comparison, encourage them to explain exactly why this procedure works—or to figure out other strategies that they can explain.

Activity

Discussing the Changes

Have the groups present their findings to the class, indicating where the biggest weight losses (or gains) were. As the students describe their findings, ask them to talk about the differences they observed:

What foods seemed to lose the most weight? the least?

Help students think about which differences were bigger. Some will notice that a large *absolute* difference isn't the same as a large *relative* difference. That is, a small cucumber slice loses a lot of weight in a day (about three-quarters of its weight), while a larger tomato slice loses more total weight, but only about one-half of its own weight. Encourage these kinds of comparisons if they come up.

Discuss the predictions you recorded on chart paper during Session 1. Encourage students to re-evaluate their predictions in light of the data they gathered. You might also consider these questions:

Why did the foods lose weight? Why did some types of food lose more weight than others? Did food seem to lose more when it was left in certain places in the classroom? Why? What do you think would happen to the weight if we left the food out for a few more days and then reweighed it?

❖ **Tip for the Linguistically Diverse Classroom** Reword discussion questions so that all students can share their thoughts. For example:

Show me places in the room where the food seemed to lose more weight. Write how much you think this food will weigh tomorrow, and two days from today.

Have all students work individually on the problem on Student Sheet 4, Who Is Closer to 100? As students are working, circulate and talk to as many as you can about their strategies for solving the problem. As you evaluate their solutions, keep the following questions in mind:

■ Did the student get an answer that is very close to 100 (for example, 101 for the first hand and 98 for the second?) If not, does it appear that the focus was mostly on getting the ones to come out as close to zero as possible while neglecting the tens, or getting the tens to come out close to 100 while neglecting the ones?

■ Does the student's strategy for solving the problem make sense? In other words, was there a good understanding of how the tens combined and how the ones combined to make a number close to 100?

Use this assessment to decide which students should spend more time playing Close to 100. Students should become fluent in knowing how far a number is from the next multiple of 10 or multiple of 100. For example, they should know that 53 is 7 away from 60 and 47 away from 100. This kind of knowledge is critical in developing strategies to solve addition and subtraction problems flexibly. During Investigation 3, students can go on playing the basic game or learn a variation with an alternate scoring method (see the **Teacher Note**, Directions for Close to 100, p. 8).

Adding with Money, Inches, and Time

What Happens

Sessions 1 and 2: Heights and Coupons
Students work on three activities in Choice Time
over the two sessions. They solve problems with
multiple addends, in which a total measurement
or total amount of money is given and they must
figure out a set of numbers to make that sum. For
example, if the total of 6 third graders' heights is
318 inches, what could each height be? If I have
$2.50, which 3 items can I buy? They also learn
the scoring variation for Close to 100, which
involves positive and negative scores. Students
discuss their addition strategies at the end of
each session.

Session 3: Planning a Party   Students plan the
activities and timing for a party that will last
exactly two hours. They brainstorm possible
activities, then work in pairs to write a party
schedule. A Teacher Checkpoint activity helps you
gauge students' progress in developing addition
strategies.

Mathematical Emphasis

■ Solving addition problems with multiple
  addends and keeping track of the steps

■ Developing a repertoire of addition strategies
  that rely on students' number sense and understanding of number relationships

■ Recognizing and using standard addition notation while using approaches based on sound
  number and operation sense

■ Exploring number relationships and using
  important equivalencies in time, money, and
  linear measure

■ Using estimation to make good approximations

What to Plan Ahead of Time

Materials

- Interlocking cubes or counters: a supply for each small group (Sessions 1–3)
- 300 charts (Sessions 1–3)
- Calculators (Sessions 1–3)
- Numeral Cards: 5 decks (Sessions 1–2)
- Yardsticks or measuring tapes marked in inches: 5–10 for the class (Sessions 1–2)
- Play money: a supply of bills and coins (Sessions 1–2, optional)
- Colored pencils, markers, or crayons (Session 3, optional)

Other Preparation

- Duplicate teaching resources and student sheets (located at the end of this unit) in the following quantities. If you have Student Activity Booklets, no copying is needed.

For Sessions 1–2

Student Sheet 1, Close to 100 Score Sheet (p. 79): 1–2 per student

Student Sheet 5, Children's Heights (p. 84): 1 per student

Student Sheet 6, Coupons Add Up (p. 85): 1 per student

Student Sheet 7, Different Ways to Add (p. 86): 1 per student (homework)

Coupons (p. 89): 1 set per pair (whole sheets, or cut apart)

Close to 100 Scoring Variation: Counting Track (p. 91): 1 per student

For Session 3

Student Sheet 8, Planning a Party (p. 87): 1 per pair

Student Sheet 9, Choose a Problem (p. 88): 1 per student

- Instead of the blackline master Coupons, you may want to substitute real coupons, which will be more interesting to your students. You could ask the class to bring in coupons from newspapers, magazines, or local stores. If you do this, you will need to revise Student Sheet 6, Coupons Add Up, so that the total amounts are possible to make with the real coupons. (Sessions 1–2)

- Try the variation of the Close to 100 game (described on p. 9) to see how the change in scoring affects your strategy for playing the game. Use the information from the assessment activity in the preceding investigation (p. 25) to decide which of your students are ready to learn this variation. (Session 1)

- Before Session 3, look over the homework from Sessions 1–2 (Student Sheet 7, Different Ways to Add), for discussion as part of the Teacher Checkpoint.

Heights and Coupons

Materials

- Interlocking cubes or counters
- Yardsticks or measuring tapes (See note on p. I-18.)
- Calculators
- 300 charts
- Numeral Card decks
- Student Sheet 1 (1–2 per student)
- Student Sheet 5 (1 per student)
- Student Sheet 6 (1 per student)
- Close to 100 Scoring Variation: Counting Track (1 per student)
- Coupons (2 pages or 1 cut-apart set per pair)
- Play coins and paper money (optional)
- Student Sheet 7 (1 per student, homework)

What Happens

Students work on three activities in Choice Time over the two sessions. They solve problems with multiple addends, in which a total measurement or total amount of money is given and they must figure out a set of numbers to make that sum. For example, if the total of 6 third graders' heights is 318 inches, what could each height be? If I have $2.50, which 3 items can I buy? They also learn the scoring variation for Close to 100, which involves positive and negative scores. Students discuss their addition strategies at the end of each session. Their work focuses on:

- estimating the sum of several addends
- using estimation to find an approximate solution that can be adjusted
- adding in the contexts of money and measurement
- developing addition strategies that make sense
- explaining addition strategies
- choosing quantities that are realistic to solve a problem

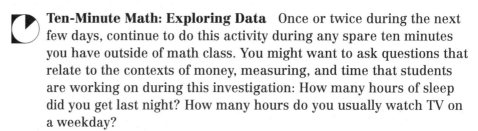

Ten-Minute Math: Exploring Data   Once or twice during the next few days, continue to do this activity during any spare ten minutes you have outside of math class. You might want to ask questions that relate to the contexts of money, measuring, and time that students are working on during this investigation: How many hours of sleep did you get last night? How many hours do you usually watch TV on a weekday?

For a later session, you can ask students to bring in data from home, for example: How tall is the tallest (shortest) person that lives in your house? How much does a quart of milk (or orange juice, or jar of peanut butter) cost at your neighborhood store?

Quickly collect and graph the data, and ask students to describe what they see. For more information about this activity, see p. 71.

Choice Time: Finding Numbers to Make a Total

Three Choices During this session and the next, students may choose from three activities that are going on simultaneously in the classroom. Students work independently most of the time on three activity choices; plan to call the class together at one point during each session for a discussion (see pp. 32–33). Some students may have time to complete only one activity and begin a second during each session.

How to Set Up the Choices If you set up the activity choices at centers, show students what they will find at each one. Otherwise, make sure students know where to get the materials needed for each activity:

Choice 1: Children's Heights—copies of Student Sheet 5, yardsticks or measuring tapes, cubes, 300 charts, and calculators

Choice 2: Coupons—sets of blackline-master coupons or real coupons, copies of Student Sheet 6, cubes, 300 charts, calculators, and play money

Choice 3: Close to 100—Numeral Card decks, copies of Student Sheet 1 and the Close to 100 Scoring Variation: Counting Track, cubes, and calculators

At the beginning of Session 1, spend a few minutes introducing Choice 1, Children's Heights, and Choice 2, Coupons, as explained in the following activity descriptions. Students have already played Close to 100 (Choice 3); introduce the scoring variation gradually as students seem ready.

Choice 1: Children's Heights

Introduce this activity by asking students if they know what their own heights are. (If your students have done the grade 3 Measuring and Data unit, *From Paces to Feet*, they may remember what their heights were earlier in the year. Of course, their heights may have changed since then!) Ask for an estimate:

If six of you were to lie down in the hall, with one person's head touching the next person's feet, how long do you think that line of people would be? Why do you think so?

❖ **Tip for the Linguistically Diverse Classroom** Help students to visualize this situation by drawing stick figures of varying heights lying head to feet in a line, with a total measurement of 318 inches.

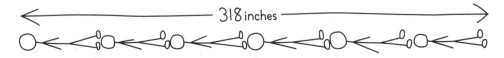

Tell students that in this activity, they will be figuring out what the heights of six third graders might be if the sum of their heights is 318 inches. Students work in pairs, but will fill out individual copies of Student Sheet 5, Children's Heights. In the first problem, students figure out any heights that are realistic for third graders and that add up to 318 inches. In the second problem, they start with two actual heights, then find four other heights to total 318 inches.

Choice 2: Coupons

Introduce this activity with a brief discussion of coupons. Students should understand that the money amount shown on a coupon will save the buyer that much money. Explain that students will be choosing coupons that add up to a certain amount of savings.

For example, suppose you wanted to find three amounts that added up to exactly $1.00. What could they be? What if you wanted to find three amounts that added up to exactly $1.50? Suppose I say you can't use a $1.00 coupon as one of your amounts?

As necessary, review how to read an amount written in standard money notation; for example, $2.50.

Working in pairs, students choose grocery coupons to fit particular criteria. For example, they are asked to select coupons for things that can't be eaten and that add up to $2.75. Students each fill out their own copy of Student Sheet 6, Coupons Add Up. If you or the students have brought in real coupons, have students use those along with the new version of Coupons Add Up that you have created to go with them. Some students may want to use play money to model the problem. Make sure the money is easily accessible. If you think some students would particularly benefit from using the money, put it within their reach.

❖ **Tip for the Linguistically Diverse Classroom** Help students understand the directions for Student Sheet 6 by reading them aloud and modeling corresponding actions—for example, by setting aside cereal coupons, making eating motions over food coupons, and shaking your head over nonfood coupons.

Choice 3: Close to 100

Students continue playing the Close to 100 game, following the basic rules they learned earlier in the unit. Gradually teach them the scoring variation described in the **Teacher Note**, Directions for Close to 100 (p. 8). The information you collected in the assessment activity (p. 25) will help you decide which students should continue to play the basic game and which students should learn the new scoring. In this version, students score +3 if they are

3 over 100, and −3 if they are 3 under 100. This scoring will change a player's strategy. The object is now to balance the scores over 100 with the scores under 100 to get a total score as close to 0 as possible. Don't expect students to understand right away how the scoring makes a difference in their strategy. Tell them to play a practice game so that they can see how the scoring works.

Students' work in the grade 3 Changes unit, *Up and Down the Number Line*, prepares them for adding up these negative and positive scores. Students who have not as yet done this work will need more help. Show them how to use the Close to 100 Scoring Variation: Counting Track to help them figure out their total scores, as explained in the **Teacher Note,** Directions for Close to 100 (p. 8).

Introduce the new scoring to groups of 6–8 students at a time. Students who understand the idea of positive and negative changes can also use the calculator to help them total their score. Once you have introduced this variation, students can break up into pairs or threes to play the game. Students who have learned the new version can also teach other students how to play.

While Students Are Working on the Choices

Observe your students at work and encourage them to use strategies that make sense to them for solving the problems. The Heights and Coupons activities (as well as the Party Planning activity they will do in Session 3) are what we call "construction problems." Instead of finding the total of several given amounts, students must work backwards from a total to find numbers that, when added together, make that total. We have found that this kind of problem pushes students to think harder about the relationship between a whole (the sum) and its parts (the addends that make that sum).

Observing students while they work on these activities can provide a great deal of insight into their understanding of relationships among quantities and operations. As you observe them, encourage students to make an estimate before they add and to use left-to-right strategies and other approaches that make sense to them for addition. Students should have 300 charts and interlocking cubes available for adding, and should always check their addition using more than one method. Methods might include using a calculator, counting on the 300 chart, using cubes put together in groups of tens, as well as mental arithmetic strategies. See the **Teacher Notes**, Three Powerful Addition Strategies (p. 37) and Using Concrete Materials for Addition and Subtraction (p. 17).

At the end of Session 1, conduct Discussion 1: Strategies for Addition (p. 32). Sometime during Session 2, pull the class together for Discussion 2, Finding Numbers to Make a Total (p. 33).

Discussion 1: Strategies for Addition

Leave 15 minutes at the end of Session 1 to discuss addition strategies. Pose a problem, using either student heights or coupons. First ask students to make an estimate of the answer; for example, is it more or less than 300 inches? Then have students discuss mental strategies or strategies using cubes or the 300 chart that they could use to figure out the result exactly. For example, use problems like the following, but substitute real students' heights or prices from real coupons you have brought in:

1. **I'm going to write down the heights of three students in the class. I'm only going to give you about 5 seconds to look at them. See if you can figure out whether the total of the heights is more than 100 inches or less than 100 inches [or 125 inches, or 150 inches]. Don't try to add them all up exactly. Just see if you can figure out whether they're *more* or *less* than 100 inches. [*Write down three actual heights.*]**

   What do you think—more or less than 100? Why do you think so?

   Now we're going to add up these heights exactly. How would you do it? Who has a way to begin?

2. **I'm going to write down the prices of three items I want to buy at the grocery store. I have $5.00 to pay for them. I'm going to give you a quick look at the prices, and I want you to estimate whether or not I have enough money to pay for them. Here are the prices: $2.07, $1.49, and $1.99.**

   Do you think the sum of these numbers is more or less than $5? Why do you think so?

   Now we're going to add these up exactly. Who has an idea about how to make this addition easier?

In this discussion and throughout their work in these sessions, encourage students to look for numbers that make 10 or 100. For example, given the problem 47 + 56 + 53:

> I know that 50 + 50 makes 100, then 40 more is 140. Then the 3 and 7 make another 10, that's 150, then 6 more is 156.

Also encourage them to find a number that is close to the number they are adding, but easier to use, then adjust the total. For example, given the problem $2.07 + $1.49 + $1.99:

> $1.49 is almost $1.50 and $1.99 is almost $2.00, so that's $3.50. Add on $2.07, that's $5.57, but I added on an extra cent to the $1.49 and an extra cent to the $1.99. So take 2 cents away; it's really $5.55.

You will find that students generally make better estimates and add more accurately when they add the larger parts of the quantities first rather than, as we were taught, adding the ones first. See the **Teacher Note**, Three Powerful Addition Strategies (p. 37).

As students tell you their strategies, record them on the board and write down intermediate steps. Tell students that when they're adding mentally, especially when they are adding more than one number, it often helps to jot down partial answers as they go along, so they don't have to keep everything in their heads. This is discussed further in the **Teacher Note**, Keeping Track of Addition and Subtraction (p. 39).

Use the words *total* and *sum* interchangeably so that students get used to hearing these two mathematical terms used correctly in context. You don't need to insist that students use them as long as they can explain their methods clearly, but they should be able to recognize what you mean when you use them.

Activity

Discussion 2: Finding Numbers to Make a Total

Sometime during Session 2, when all students have had a chance to finish either the Children's Heights activity or a few of the Coupons problems, stop everyone to have a brief discussion about how they are finding numbers to make a given sum:

What strategies are you using to find numbers to add together to make a certain total? How did you start? Do you remember times you got stuck?

After students contribute some observations, pose the following question, jotting down the amounts on the board as you give them:

I went into a bookstore that was having a sale on books. I had $2.50 and I wanted to buy three different books. The books had lots of different prices, from around 50 cents up to $5.00. I picked out one book that cost 99 cents and another one that cost $1.25. Do you think these are good choices so far? Will I have enough money left to buy one more book? How do you know?

❖ **Tip for the Linguistically Diverse Classroom** You might act out this problem to make it more comprehensible, using books on a table with sale prices marked on stick-on notes.

To help all students participate, give them a few minutes to talk with a partner about the problem before they volunteer their opinions.

Do one or two more problems like this one. You might pose a Coupons problem, using some real coupons. For example, write the amounts and items from six or seven coupons on the board. Ask:

Who can find three coupons that add up to $3.00? How would you start? Which coupons seem like they might work? Which ones do you know right away wouldn't be good choices? Why?

Sessions 1 and 2 Follow-Up

 **Homework**

Different Ways to Add   After Session 1, send home Student Sheet 7, Different Ways to Add. Students think of three ways to solve each addition problem and find some way to record their approaches so that someone else can understand their strategies. Using the calculator can be one of their ways, but they still must record how they used it. Remember that any of the strategies described in the **Teacher Note**, Three Powerful Addition Strategies (p. 37), can also be used on the calculator. Remind students to bring this homework back to class so that you can collect and look over the sheets before Session 3. Then you can ask students about any strategies you don't understand during the Teacher Checkpoint in that session.

 **Extensions**

Inches and Feet   The Children's Heights activity offers an opportunity to review the relationship between inches and feet. Small groups of students can add together their heights in inches, then determine how many feet the total is close to. Review with students that there are 12 inches in a foot. As the students help you count by 12's, record their count where you can leave it posted. Record the information as: 1 foot = 12 inches, 2 feet = 24 inches, 3 feet = 36 inches, and so on. Use this list to ask such questions as:

About how many feet is 318 inches?

If I am 62 inches tall, about how many feet tall am I?

Yesterday Latisha measured the doorway. It was 76 inches; about how many feet is that?

Measuring in Centimeters   Students can try the heights problems on Student Sheet 5 using centimeters. Substitute 810 centimeters for the total height. Students will be interested in the larger numbers they are getting for their measurements. Ask why they think the numbers are so much larger.

Planning a Party

What Happens

Students plan the activities and timing for a party that will last exactly two hours. They brainstorm possible activities, then work in pairs to write a party schedule. A Teacher Checkpoint activity helps you gauge students' progress in developing addition strategies. Student work focuses on:

- working with multiple addends in the context of time
- developing addition strategies that make sense
- explaining their addition strategies
- choosing quantities that are realistic to solve a problem

Materials

- Student Sheet 8 (1 per pair)
- Student Sheet 9 (1 per student)
- Cubes or counters, 300 charts
- Calculators
- Colored pencils, markers, or crayons (optional)

Planning a Party

Tell students that they will be planning a party that starts at exactly 5:00 P.M. and ends at exactly 7:00 P.M. (Don't tell the class that this is two hours.) Explain that they have to plan carefully which activities they want to do and how much time each will need. As a group, students brainstorm the kinds of activities they like to do at parties. List all ideas on the board. They can make this a particular kind of party, such as a birthday party or a seasonal celebration, if they wish.

Students work in pairs with one copy of Student Sheet 8, Planning a Party, to fill out jointly. Each pair can choose activities from those listed on the board to include in their party, or can use their own ideas.

❖ **Tip for the Linguistically Diverse Classroom** Pair English-proficient students with those who have limited English proficiency. Offer the option of using words, drawings, or both when listing the party activities.

This task can be quite difficult. Students do not automatically know that the party will be two hours, or that two hours is the same as 120 minutes. Let them figure this out with their partners as they work on the problem. Some students might enjoy drawing a picture of their party and posting this along with their activity schedule in the classroom.

Teacher Checkpoint

Addition Strategies

At the end of this session, ask students to solve one of the problems on Student Sheet 9, Choose a Problem, using two or three different strategies.

One way may be by using the calculator, cubes, or 300 chart, and at least one should be a mental strategy. Help students choose a problem that is sufficiently challenging, but not too difficult for them.

While students are working, talk with them about how they are solving the problem. Spend time with students who might be "falling through the cracks"—those you have not worked with recently and are not sure about how they are doing. You can also use this time to return the homework on Student Sheet 7, Different Ways to Add, and ask students to explain anything you didn't understand or to correct solutions you didn't agree with. As you observe students' work, consider the following questions:

- Are students developing their own strategies?
- Are the strategies efficient, reliable?
- Are they able to jot down intermediate steps to help them keep track of their procedures?
- Do they have good ways of double-checking their results?

Session 3 Follow-Up

Homework

Choose a Problem　Students solve two or three more problems on Student Sheet 9, Choose a Problem.

Extension

Planning a Class Activity　Students would, of course, enjoy planning a real class party or a class field trip. If you have an upcoming trip planned, or can plan a walking trip to a nearby park, playground, or other neighborhood attraction, have students help plan the time and activities of the trip. Use real time constraints for when the class will leave the school building and when they must be back. For example, to plan a trip to the park, they will need to think about the time it takes to walk to the park, several activities at the park (perhaps one all-class game, individual play time, time for you to read aloud a book to the class, and snack time), and the time to walk back to school.

Three Powerful Addition Strategies

Most of us who are teaching today learned to add starting with the ones, then the tens, then the hundreds, and so on, moving from right to left and "carrying" from one column to another. This algorithm is certainly efficient once it's mastered. However, there are many other ways of adding that are just as efficient, closer to how we naturally think about quantities, connect better with good estimation strategies, and generally result in fewer errors.

When students rely only on memorized rules and procedures that they do not understand, they usually do not estimate or double-check. They may get results that make no sense, considering the numbers involved in the problems. We want students to use strategies that encourage, rather than discourage, them to think about the quantities they are using and what to expect as the result. We want them to use their knowledge of the number system and important landmarks in that system. We want them to easily break apart and recombine numbers in ways that help them make computation more straightforward and, therefore, less prone to error.

The three powerful addition strategies discussed here are familiar to many competent users of mathematics. Your students may well invent others. It is critical that every student be comfortable with more than one way of adding so that an answer obtained using one method can be checked by using another. Anyone can make a mistake while doing routine computation—even with a calculator. What is critical, when accuracy matters, is that you have spent enough time estimating and double-checking to be able to rely on your result.

Left-to-Right Addition: Biggest Quantities First When students develop their own strategies for addition from an early age, they usually move from left to right, starting with the bigger parts of the quantities. For example, when adding 47 + 48, a student might say "40 and 40 is 80, then 8 and 7 is 15, so 80 plus 10 more is

90 and then 5 more makes 95." This strategy is both efficient and accurate. Some people who are extremely good at computation use this strategy as their basic approach to addition, even with large numbers.

One advantage of this approach for students is that when they work with the largest quantities first, it's easier to maintain a good sense of what the final sum should be. Another advantage is that students keep seeing the quantities 47 and 48 as whole quantities, rather than breaking them up into their separate digits and losing track of the whole. When using the traditional algorithm (8 + 7 = 15, put down the 5, carry the 1), students too often see the 8, the 7, the 5, the 1, and the two 4's as individual digits. They lose their sense of the quantities involved and if they end up with a nonsensical answer, they do not see it because they "did it the right way."

Rounding to Nearby Landmarks Changing a number to a more familiar one that is easier to compute is another strategy that students should develop. Multiples of 10 and multiples of 100 are especially useful landmarks for students at this age. For example, in order to add 199 and 149, you might think of the problem as 200 plus 150, find the total of 350, then subtract 2 to compensate for the 2 added on at the beginning.

Of course there are other useful landmarks, too. If you are adding 23, 26, and 27, you might use 25 as your landmark, rather than 20 or 30:

> Three 25's would be 75, so I'm 2 under and 2 over with the 23 and 27, so just add 1 more from the 26, and it's 76.

There are no rules about which landmarks in the number system are best. It simply depends on whether using landmarks helps you solve the problem.

Changing the Order of the Numbers Simply changing the order of the numbers you are adding is often a great help. For example, when

Continued on next page

adding 23 + 46 + 7, the problem becomes much simpler as soon as you recognize that 23 + 7 is 30. Changing the order of numbers can also involve partitioning some numbers into two parts. For example, if you are adding 108 + 45 + 162, you might add this way:

> 160 plus 40 is 200, plus another 100 is 300; 2 and 8 is 10 plus 5 is 15, so it's 315.

These strategies may be used alone or in combination, whether the problem is being done mentally, on paper, or with the calculator. Encourage students to get into the habit of always looking over the whole problem before they begin solving it. Are there numbers they can combine easily? Are there useful landmark numbers they can use? Will they solve it by adding from left to right?

There are no hard and fast rules about which strategies are best for which problems. It really depends on what works for a particular person and how that person sees a particular problem. Even when you may think that a particular strategy is clearly best, students can surprise you in

the way that they see the problem. For example, we may think that it is obvious to change 199 + 149 to 200 + 150, then subtract 2. However, someone else might use the following method, just as efficient and accurate:

> 199 plus 100 is 299, then I'll take 1 from the 49 to make 300, leaving 48, so it's 348.

Similarly, while one person might use 25 as a landmark to solve 23 + 26 + 27, another might rearrange the numbers in the problem, adding 23 + 27 first to get 50, then adding on the 26.

If you have students who have already memorized the traditional right-to-left algorithm and believe that this is how they are "supposed" to do addition, you will have to work hard to instill some new values—that estimating the result is critical, that having more than one strategy is a necessary part of doing computation, and that using what you know about the numbers to simplify the problem leads to procedures that make more sense, and are therefore used more accurately.

Keeping Track of Addition and Subtraction

During mental computation, especially when several numbers are involved, students should get into the habit of jotting down intermediate steps so that they don't lose track of their procedures. For example, suppose you are adding the four numbers below. You might jot down the partial sums like this as you add from left to right:

1540	3000 [the sum of the 1000 and 2000]
347	1700 [the sum of 500, 300, and 900]
2063	150 [the sum of 40, 40, 60, and 10]
+ 918	18 [the sum of 7, 3, and 8]

Now you can easily add the partial sums in your head and write down the result.

Encourage students to jot down their intermediate steps. Show them what you mean by recording for them as they tell you their strategies. Then, insist that they jot down their steps in order to remember their approaches and explain them. For example, Jennifer is adding up several prices. Here is what she might think and write:

$2.07 + $1.49 + $1.99

Jennifer thinks:	She records:
I'll make $1.49 into $1.50 and $1.99 into $2.00	$3.50
Then 2 more dollars from the $2.07	$5.50
Add on the 7 cents from the $2.07	$5.57
Subtract the 2 extra cents I put on at the beginning	$5.55

Keeping track of these intermediate steps helps Jennifer organize her thinking and provides a record so that she can later explain her reasoning.

For problems of comparison, a number line notation will make sense to some students. For example, suppose the class is comparing right and left handfuls of beans (as they do in Investigation 4).

> I have 84 beans in my right hand and 142 in my left. How many more could I hold in my left hand?

A student could write:

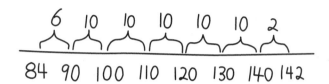

Others might be more comfortable jotting down words and numbers in a list:

84 to 90	6
90 to 100	10
100 to 140	40
140 to 142	2

Still others might use phrases:

> 84, 94, 104, 114, 124, 134, 144 makes 60.
> Take 2 away, it's 58.

Students need to be told that "taking notes" while they are working on their own procedures is acceptable, valued, and necessary. Otherwise, students often hide their jottings or try to keep everything in their heads because they think they should only write down answers or certain accepted procedures. When you do problems with the whole class, demonstrate different ways of recording what you are doing and solicit other ideas from students.

Working with Hundreds

What Happens

Session 1: Handfuls of Beans   Students compare the number of beans they can grab in their right and left hands and find the difference between the two quantities. All the data are collected; then students consider the data for "right handfuls," first in a list, then on a line plot. For homework, they collect more data about how much a hand can hold.

Session 2: More Handfuls   Students share the results of their homework. They then solve two handfuls problems and discuss how standard addition and subtraction notation can be used to record these problems and their solutions. As an assessment activity, students work on two problems based on the handfuls data collected in the previous session, write about their approach to solving the problems, and record their solutions using standard notation.

Sessions 3 and 4: Hundreds of Paper Clips
Students use boxes of paper clips to represent groups of 100. They combine hundreds into thousands and work on recognizing the words and numerals for these numbers. During Choice Time, students work on Paper Clip Problems, in which they subtract small amounts from multiples of 100; write their own Paper Clip Problems; and explore Related Problem Sets—groups of related problems in which the solution to one problem is used to help solve others.

Mathematical Emphasis

- Developing and communicating strategies for combining and comparing quantities in the hundreds and thousands
- Using standard addition and subtraction notation to record comparison problems
- Using multiples of 100 as landmarks for adding and subtracting
- Collecting, recording, and graphing data
- Making predictions about data, describing and interpreting data

What to Plan Ahead of Time

Materials

- Containers of dried beans: 1 per small group (Session 1). The container needs to be large enough for students to put one hand in and withdraw a handful of beans. Red beans or kidney beans are about the right size to give you some numbers in the hundreds in your data.
- Cubes and 300 charts remain available (Sessions 2–4)
- Boxes of paper clips (100 per box): 1 per student (Sessions 3–4). You won't actually use any of these clips, so they could be borrowed and returned to the school supply room. If you can't get paper clips, use any small item in groups of 100, for example, 100 beans in a plastic bag. Another simple solution is for each student to use a 100 chart, with the squares representing 100 things.
- Overhead projector (Sessions 3–4)
- Calculators (Sessions 3–4)
- Colored pencils, markers, or crayons (Sessions 3–4)
- Envelopes for storing Paper Clip Problems and Related Problem Sets (Sessions 3–4)
- Paste or glue sticks (Sessions 3–4)

Other Preparation

- Duplicate student sheets and teaching resources (located at the end of this unit) in the following quantities. If you have Student Activity Booklets, copy only the transparency needed.

For Session 1

Student Sheet 10, Handfuls at Home (p. 92): 1 per student (homework)

For Session 2

Student Sheet 11, Handfuls of Beans (p. 93): 1 per student (optional)

Student Sheet 12, Making Comparisons (p. 94): 1 per student

Student Sheet 13, Class Handfuls and My Handfuls (p. 95): 1 per student (homework)

For Sessions 3–4

Quick Image Paper Clip Boxes* (p. 97): 1 transparency, cut apart

Paper Clip Problems (p. 98): 1 per student and 1 extra. Cut sets apart into individual problems. Store the copies of each problem in a separate envelope. Paste an example of each problem on the front of the envelope so students can see which problem they are choosing.

Related Problem Sets (p. 99): 1 per student and 1 extra. Cut the sheet apart into individual problem sets. Store the copies of each problem set in a separate envelope. Paste an example of each problem set on the front of the envelope so students can see which problem set they are choosing.

Student Sheet 14, Problem Strategies (p. 96): 1 per student (homework)

- Collect your own data about handfuls of beans. Grab a handful of beans with your right hand, count them, and record the number. Do the same with your left hand. (Session 1)

Handfuls of Beans

Materials

- Containers of beans (1 per 3–4 students)
- Student Sheet 10 (1 per student, homework)

What Happens

Students compare the number of beans they can grab in their right and left hands and find the difference between the two quantities. All the data are collected; then students consider the data for "right handfuls," first in a list, then on a line plot. For homework, they collect more data about how much a hand can hold. Their work focuses on:

- developing sound strategies for comparing two numbers
- organizing a counting task to ensure accuracy
- making predictions
- collecting, displaying, and interpreting data
- using a line plot

 Ten-Minute Math: Estimation and Number Sense Once or twice during the next few days, outside of math time, do this activity with a particular focus on multiple addends.

Use the board or overhead to present a problem with multiple addends. For example:

$$30 + 54 + 30$$

$$\$2.15 + \$3.00 + \$.95$$

20 minutes plus 55 minutes plus 1 hour

Allow students to view the problem for a short time—maybe 10 to 15 seconds. Ask them to come up with the best estimate they can. Tell them not to write anything down or use the calculator during this time.

Cover up the problem and have students discuss what they know. Encourage all kinds of estimation strategies as well as precise answers:

It's definitely bigger than 100 because there was a 50, and then a 30 and 30, which is more than another 50.

It has to be 2 hours and 15 minutes if you take 5 from the 20 and add it to the 55.

Uncover the problem and continue the discussion. For variations on this activity and ways to integrate the calculator, see p. 73.

Comparing Handfuls of Beans

Show students the containers of beans. If you have already worked through the introductory unit, *Mathematical Thinking at Grade 3*, students will be familiar with this activity. You might point out to them that these beans are smaller objects than they used in that unit.

How many beans do you think you can grab out of this container and hold with one hand?

Ask volunteers for estimates. Then demonstrate how to grab the beans and have someone come up and grab a handful. At this point, you might want to discuss some rules for grabbing beans. For example, should you be allowed to scoop up the beans so they are piled high on your hand? Should you count the dropped beans? As questions come up, the class decides on appropriate rules. You might have a couple of volunteers demonstrate how to grab the beans according to the class rules.

We're going to compare how many beans you can hold in your right hand with how many you can hold in your left hand. Which hand do you think can hold more, your right or your left? Or would they hold about the same?

Get students' reasons for their predictions.

Teacher's Bean Data   For this part of the demonstration, use the data that you collected yourself before class. You could have a student demonstrate if you prefer, but remember that it takes quite a while to count the beans. Tell students about your data:

Last night I tried this myself. I grabbed 127 beans with my right hand and 78 beans with my left hand.

Write both numbers on the board and ask students to estimate the difference:

My right hand	My left hand
127 beans	78 beans

Without figuring it out exactly, about how many more beans did I hold in my right hand? About what was the difference—close to 10 more, 50 more, 100 more? How do you know?

Ask students to share some strategies for finding the difference between the two numbers. Emphasize that figuring out the solution in more than one way is important so that you have double-checked the answer. Ask students to identify some landmark numbers that might help them solve the problem.

Students' Bean Data   After discussing their strategies, set out the containers of beans and have students do their own comparison of right and left hands. They might want to do this a couple of times.

Emphasize that students are responsible for carefully counting and recording their data. As you circulate around the classroom, help students organize their counting strategies so that they can be sure to count accurately: Will groups of 2's, 5's, 10's, or 20's help them organize their beans? How can they double-check their counting?

Students then work together to figure out how many more beans were held in one hand than the other. They might use the beans themselves, a 300 chart, or interlocking cubes to prove their solution. Some students may have an easier time visualizing how to solve this problem if the question is formulated this way:

How many beans would you need to add to your left hand so it would have the same amount as the right hand?

As you circulate, ask students to show you how they can prove their solution. Urge students to jot down the way they solved the problem, using lists, phrases, or pictures to keep track of what they are doing and explain it clearly to you or someone else. See the **Teacher Note**, Keeping Track of Addition and Subtraction (p. 39), for different ways to keep track.

Record the number of beans each student held in right and left hands and the difference between them. You may want to list these numbers on the board or overhead without identifying students' names in order to decrease any competitiveness about who can hold more beans. List the numbers as you get them from students or have students add their own numbers to the list, but don't put them in order at this point.

Right hand	Left hand	Difference
85	79	6
92	100	8
78	80	2
82	70	12
70	78	8

**Looking at the
Handfuls Data**

What do you notice about the data we collected?

After students make a few observations, ask them if there is an easier way
to look at the data. Students might suggest putting the data in an ordered
list or graph. Introduce the line plot as one quick way to see the data in
order:

**I'm going to put these on a graph called a line plot. It looks like a num-
ber line. We're going to put up the data for just the *right* handfuls first.
What's the smallest handful we have? What about the largest? I don't
want to make this line plot by 1's—it would be too long. What should I
use?**

If students have worked with the grade 3 Measuring and Data unit, *From
Paces to Feet*, they will be familiar with line plots. If not, introduce this kind
of graph now. See the **Teacher Note**, Line Plot: A Quick Way to Show the
Shape of the Data (p. 47), for more information.

Get students to help you decide what the beginning and ending numbers
for your line plot should be and what number you should count by. Then
put the students' data on the plot. Students can help by reading the data
out loud, or you can go around the room and have students say their own
piece of data. Your line plot might look something like this:

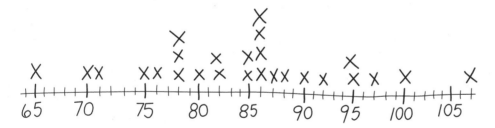

After the line plot is made, tell students to look again at the data and ask:

What do you notice about the data from looking at the graph?

Encourage students to make observations about the range of the number
of beans, the minimum and maximum numbers, whether there are any
unusual numbers, whether the numbers are clumped together or spread
out. You might also ask:

If someone asked you about how many beans a third grader can hold, what would you say?

What's the middle-sized handful?

If we added the data for the left hand in a different color, would it be in about the same places on the graph, or would it be different?

Note: Keep the line plot and list of handfuls data posted for students to use during the assessment in Session 2.

Session 1 Follow-Up

 **Homework**

Handfuls at Home   Send home copies of Student Sheet 10, Handfuls at Home. Students find something at home to use for another handfuls comparison and involve people in their families.

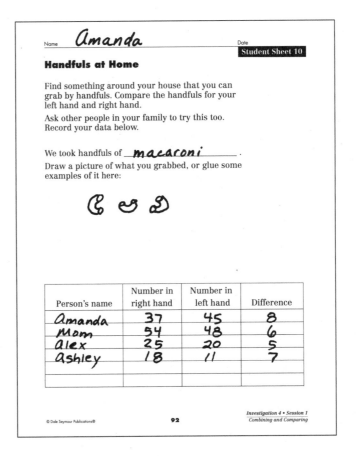

Person's name	Number in right hand	Number in left hand	Difference
Amanda	37	45	8
Mom	54	48	6
Alex	25	20	5
Ashley	18	11	7

 **Extension**

Adding the Left Hand Data   Graph the left handfuls data on another line plot, using the same scale as the one for the right hand. Place one below the other and compare the two line plots. What is the same about them? What is different?

Line Plot: A Quick Way to Show the Shape of the Data

An important part of statistics is organizing and representing data so that it is easy to see and describe. A *line plot* is one quick way to organize numerical data. It clearly shows the range of the data and how the data are distributed over that range. Line plots work especially well for numerical data with a small range.

A line plot is often used as a working graph during data analysis. That is, it is an organizing tool we can use as we begin work with a data set, not a careful, formal picture we use to present the data to someone else. Therefore, it need not include a title, labels, or a vertical axis. A line plot is simply a sketch showing the values of the data along a horizontal axis and X's to mark the frequency of those values in the data set.

For example, if 25 students have collected data on the number of beans they can grab in their right hands, a line plot showing these data might look like the one below.

From this display, we can quickly see that, while the *range* of the data is from 70 to 120, the *interval* in which most data fall is from 70 to 100, with *outliers* at 112 and 120. There is a large clump of data between 75 and 86, accounting for over half of the students' handfuls. (*Range, interval,* and *outlier* are terms that you will use with students as the need arises—introducing them informally in the context of discussing their data, rather than with formal definitions.)

One advantage of a line plot is that we can record each piece of data directly as we collect it. To set up a line plot, start with an initial guess from students about what the range of the data is likely to be: What do you think the lowest number should be? How high should we go? Leave some room on each end of the line plot so that you can lengthen the line later if the range includes lower or higher values than you expected.

By quickly sketching data in line plots on the chalkboard, you provide a model of how such plots can provide a quick, clear picture of the shape of the data.

```
                    X
    X         X  X  X        X
    X  X      X  XXXX  X  XXX        X      X X      X      X              X              X
  ┬─────────┬─────────┬─────────┬─────────┬─────────┬─────────┬─────────┬─────────┬─────────┬─────────┬─────────
  70        75        80        85        90        95        100       105       110       115       120
```

More Handfuls

What Happens

Students share the results of their homework. They then solve two handfuls problems and discuss how standard addition and subtraction notation can be used to record these problems and their solutions. As an assessment activity, students work on two problems based on the handfuls data collected in the previous session, write about their approach to solving the problems, and record their solutions using standard notation. Their work focuses on:

- developing good strategies to solve comparisons problems
- relating standard addition and subtraction notation to comparison situations
- describing their strategies in writing

Materials

- Completed Student Sheet 10 (from Session 1 homework)
- Student Sheet 11: 1 per student (optional)
- Student Sheet 12: 1 per student
- Student Sheet 13: 1 per student (homework)
- Cubes and 300 charts

Activity

Handfuls from Home

Ask students to share the kinds of things they made handfuls of at home. Ask other students to estimate about how large a handful of each object might be.

Midori counted a handful of rice. Don't tell us how much was in it yet, but I'd like to know, how did you count it? Did it take a long time? Did you have any strategies for counting accurately?

Now, does anyone have an estimate of how many grains of rice Midori might have been able to hold in her hand?

You can also ask about how students' own handfuls compared with others in their families:

Did anyone find someone at home with a very big or very small hand? What did they hold? How many could they hold compared to how many you could hold?

Sean found out that his little sister could hold 27 popcorn kernels and he could hold 96. About how many more do you think Sean had in his hand?

Recording Comparisons with Standard Notation

This activity presents two problems, which you can provide to students on Student Sheet 11, Handfuls of Beans. Alternatively, you might simply present the two problems at the board or overhead. If you are not using the student sheet, explain the first problem:

Let's say I grabbed 189 beans with my right hand and 150 beans with my left hand. What's the difference between the number of beans in my two hands?

So that students can see the numbers in the problem, you might want to write them on the board or overhead, but don't write the problem using addition and subtraction notation yet. Simply record the information:

> Right hand—189 beans   Left hand—150 beans

Ask students to work on this problem for a few minutes and to write down their strategy for solving it. They may work with a partner if they wish. When students are satisfied with their solutions, have a few of them share their strategies.

Using Standard Notation   Now show students the standard ways to write the problem and its solution using addition and subtraction notation:

$$189 - 150 = 39 \qquad 150 + 39 = 189$$

$$\begin{array}{r} 189 \\ -\,150 \\ \hline 39 \end{array} \qquad \begin{array}{r} 150 \\ +\,39 \\ \hline 189 \end{array}$$

Some people think about this as addition: 150 beans in my left hand plus 39 more beans makes 189 beans in my right hand. Other people think about this problem as subtraction—189 beans in my right hand, compared with 150 beans in my left hand, and the difference is 39 beans.

All the notations above show the same situation (see the **Teacher Note**, Adding and Subtracting Go Together, p. 52). Ask students to say what each number problem means in words, describing them in terms of the beans. Thus, for example, if a student says, "189 − 150 = 39 means 189 take away 150 equals 39," ask how this works with the beans. A student might respond:

> If I start with the 189 beans in my right hand, and I take away 39 beans from that pile, I get 150 beans—the number I have in my left hand.

Encourage language that uses the words *comparison* and *difference*. Many students may not see comparison problems as subtraction, so you will need to help students read standard notation in this way.

Students work in pairs on the second problem. If you are not using Student Sheet 11, write the problem on the board or overhead:

> Suppose you can hold 150 beans in your right hand and 217 beans in your left hand. How many more beans are in your left hand? Write down how you figured this out.

Be sure that they solve the problem using their own strategies first, rather than immediately writing it in standard notation. While it is important for students to read and use standard notation comfortably, it is also critical that the notation not interfere with students' use of their own sensible strategies, as explained in the **Teacher Note**, Developing Strategies for Addition and Subtraction (p. 16).

Activity

Assessment

Handfuls Problems

Students use Student Sheet 12, Making Comparisons, along with the posted list and line plot of your class data from the last session, as they do more comparison problems and record them in two ways. First, they write about or show the strategy they used to solve the problems. Once they have explained their strategy, they write the problem and its solution using standard addition and subtraction notation.

As they record their own strategies, ask students to draw or write clearly so that someone else who isn't in the classroom today, such as someone in their family, could understand both the problem and its solution. They don't have to use complete sentences, but others need to be able to follow what they did. This is, of course, somewhat different from just keeping notes for yourself. Students can start with their own way of jotting down intermediate steps in a list or with phrases, but they may have to add some words or pictures to make it clear to someone else.

Students can work together to solve problems, with each student recording on his or her own student sheet. Check to see that students are still using sensible strategies involving landmarks in the number system, even though they are also writing the results using standard notation.

❖ **Tip for the Linguistically Diverse Classroom** To ensure that students record the line plot data correctly, do the top part of Student Sheet 12 orally with the entire class. Then check individually with students who have limited English proficiency on problems 1 and 2.

As you observe students and review their work later, consider the following:

- Can the students solve a comparison problem with numbers of this size? Do they use landmarks in the number system, such as multiples of 10, when finding the differences?

- Do the students use materials appropriately, such as interlocking cubes or the 300 chart, to find the solution?

- Can students prove their solution to you? Do they double-check, using another strategy?

- Can the students communicate clearly about how they solved the problem?

- Do they understand how to record the problem and its solution using standard addition and subtraction notation? Does the standard notation seem to trigger the use of rote procedures rather than approaches that reflect a good understanding of number?

Session 2 Follow-Up

Class Handfuls and My Handfuls   Give a copy of Student Sheet 13, Class Handfuls and My Handfuls, to each student. Make sure students fill in the information at the top of the sheets from the class data list before they go home.

 **Homework**

❖ **Tip for the Linguistically Diverse Classroom**   Let students use words from their native language to explain how they solved each problem. Or, they use symbolic drawings and numbers.

Ordering Handfuls   Students may want to compare the number of objects of different sizes that they can grab in a handful. Make available several different-sized objects for them to grab by handfuls, such as paper clips, macaroni, or small erasers. Students order the objects in a chart according to how many of them fit into a handful.

 **Extensions**

Encourage students to think about whether more small things than large things can fit into a handful and why. Students can choose pairs of objects for which they think one will have twice as many in a handful as the other (for example, paper clips compared to erasers). They test their predictions by grabbing and counting handfuls of objects to see if the relationship "twice as many" is appropriate.

Giant Handfuls   Some students may want to try figuring out the difference between "weird" or giant handfuls. Examples might be 998 and 98, 180 and 350, 525 and 375, and 401 and 97.

An animated discussion recently took place among the group of writers on *Investigations* curriculum. "I never subtract," claimed one of the senior writers. The rest of us were shocked by this revealing statement from an excellent mathematician. "What *do* you do?" we asked. It turns out that she almost always deals with subtraction problems by adding on to the smaller number in order to reach the larger number. And, of course, this method works just fine.

In this unit, we encourage students to use either addition or subtraction, or a combination of both, to make comparisons. For example, consider how flexible 8-year-old Elena is in her use of addition and subtraction to compare 51 and 22 beans. She explained:

> I know that 30 – 22 is 8.
>
> So then add a 10 to get 40.
>
> 40 – 22 is 18.
>
> Now I just have to get to 51.
>
> 40 plus 10 is 50.
>
> So 10 more than 18 is 28.
>
> One more to go. 50 plus 1 is 51.
>
> And 28 plus 1 is 29.

Elena used an ingenious combination of addition and subtraction to compare the two numbers. Notice how well she uses resting places or landmarks in the number system (30, 40, and 50). She adds or subtracts to get to these places, then combines the results of her operations quite effectively.

As you work with all of the problems in this unit, encourage the flexible use of addition and subtraction. Ultimately, we hope students will see not only that both processes work, but also that these processes are quite related and are, in fact, the inverse of each other. Encourage students to articulate their ideas about adding, subtracting, and the relationship between them.

We overheard one third grade teacher talking with a student who had tried adding, then subtracting two numbers she was comparing—6 and 4.

When you're comparing two numbers, what are you trying to find?

Maya: You're trying to find how many more one is than the other.

When you're adding, what do you find out about the numbers?

Maya: I don't know.

Are you comparing the two numbers you're adding?

Maya: You can be. If you take a 6 and another number … No. You're combining.

Okay, so what happens when you compare them?

Maya: 6 is 2 more than 4.

How did you get that?

Maya: I added 2 onto 4.

Right, you added to compare, but did you add the two numbers you started with?

Maya: No, I didn't, that would be combining, but I need to be comparing right now.

Hundreds of Paper Clips

What Happens

Students use boxes of paper clips to represent groups of 100. They combine hundreds into thousands and work on recognizing the words and numerals for these numbers. During Choice Time, students work on Paper Clip Problems, in which they subtract small amounts from multiples of 100; write their own Paper Clip Problems; and explore Related Problem Sets—groups of related problems in which the solution to one problem is used to help solve others. Student work focuses on:

- combining hundreds
- reading and writing hundreds and thousands
- adding to and subtracting from multiples of 100

◔ **Ten-Minute Math: Estimation and Number Sense** Continue doing this activity during any spare ten minutes you have outside of mathematics class. Use subtraction problems, such as 94 – 36 or 127 – 83. Write the problems in standard notation, both vertically and horizontally, so that students get used to interpreting both forms. You may also want to put the problems in contexts that students have used in this unit: bean handfuls, weights, or ages.

Emphasize use of landmarks, such as multiples of 10 (36 to 40 is 4, 40 to 90 is 50, 90 to 94 is 4), as well as counting by 10's (46, 56, 66, 76, 86, 96, that's 60, go back 2 is 58) as students solve these problems.

For more information on this activity, see p. 73.

Materials

- Boxes of paper clips or alternative material in 100's (1 per student)
- Cubes and 300 charts
- Calculators
- Overhead projector
- Quick Image Paper Clip Boxes (transparency)
- Envelopes or resealable plastic bags for storing student Paper Clip Problems and Related Problem Sets
- Prepared envelopes or resealable plastic bags containing Paper Clip Problems and Related Problem Sets, already cut apart for classroom use
- Colored pencils, markers, or crayons
- Paste or glue sticks
- Student Sheet 14 (1 per student, homework)

Quick Images: How Many Paper Clips?

Hold up a box of paper clips. Show students the side of the box where it says there are 100 paper clips in a box. Then explain:

I'm going to show some groups of paper clip boxes on the overhead. Your job is to figure out how many paper clips there are in each group of boxes. I'll show the picture of each group for only a few seconds. Then you make an estimate of how many paper clips are in the group.

After you describe what you saw, I'll show you the picture again to check your memory. Remember that these boxes are full, and that each box always has 100 paper clips.

Show the pictures from the transparency, Quick Image Paper Clip Boxes, one at a time, starting with the group of three boxes. Show each image for about three seconds, then have students write down what they saw and how many paper clips they think are in the display.

Afterward, ask students to talk about how the boxes were arranged, the number of boxes shown, and what this means about the total number of paper clips. After students have had a chance to describe and discuss what they saw, show the image again for students to check their memories. Again, ask them to describe what they see this time. Have a student come up and show how he or she figured out how many clips there are in the display.

As the number of boxes in the display gets larger, the class will need to talk about how to read and write numbers in the hundreds. For each display, ask students how to write the number and how they would say it. If someone says "ten hundred" for a thousand, acknowledge that there are ten hundreds and ask if anyone knows the special name for ten hundreds. As you move up to higher numbers in the next activity, encourage the use of both "seventeen hundred" and "one thousand seven hundred."

Combining, Then Subtracting

Counting by 100's Pass out one box of paper clips (or other group of 100 items, or copies of 100 charts) to each student. Remind them that there are 100 clips in each box (or 100 squares on the 100 chart). Ask some questions that involve combining the boxes, using students and groups in your own class to make these questions interesting. Try to use some small combinations and some larger combinations. For example:

- If Khanh, Laurie Jo, Aaron, and Saloni put their boxes of clips together, how many would they have?
- If all the students who wear glasses put their boxes of clips together, how many would they have?
- How many paper clips do all the girls in the class have in their boxes?
- If all the boys in the class combined their boxes, how many would they have?
- How many clips do all the boxes have?

With some of these questions, students will combine hundreds until they move into the thousands. Encourage them to count up by 100 to get to the appropriate number and to name these numbers in more than one way.

Subtracting from Hundreds   Repeat one of the problems that you have just been working on that resulted in a small number of hundreds—300, 400, or 500 works well. Tell students to listen carefully to the next step in this problem, which is to subtract a small number of paper clips from the total.

We've figured out that the total number of clips is 300. Now imagine that I need to use some of those clips. I need 3 of the paper clips to put some papers together. [*Take 3 clips out of one of the boxes.*] **With your partner, figure out how many paper clips there are now.**

Give students time to work on this problem in pairs. Encourage them to use any strategies they can to figure out how many are left. Make sure that they have counting materials and calculators. They should write down their numbers and keep notes on the method they used.

Discuss what students found and how they found it. Encourage students to think about what 1 less than 300 is, then 2 less than 300, and 3 less than 300. Counting backwards by 1's, either on paper or with a calculator, is a very useful strategy for this problem.

Depending on how your students do with this problem, do one or more similar problems with the group. Make sure that students see several methods their classmates have devised, or demonstrate them yourself. For example:

- Put together three 100 charts. Mark off 3 from the last one. Now you can see there are 100 plus 100 plus 97 (from the last chart).
- Put together 300 interlocking cubes. (This is not too time-consuming if the cubes are stored in tens.) Take 3 away. Count up how many are left.
- Count backwards mentally from 300. Take one away, that's 299. Take another away, that's 298. Take another away, that's 297.
- Count backwards on the calculator. Put 300 on the calculator. Subtract 1. Subtract 1. Subtract 1. Then press the equals sign.

Some students understand the principle of subtracting, but are unsure of the number sequence when they are counting backwards. Using the calculator as described in the last example will help them become familiar with how the backwards number sequence looks.

Choice Time: Working with Hundreds

Three Choices   During the rest of Sessions 3 and 4, students choose from three activities that are going on simultaneously in the classroom. Students should try each activity sometime during the two sessions, but they need not complete all the problems. They may go back to an activity during Session 4, but they should try different problems the second time through. You may also want to assign one or two problems for homework after Session 3.

As you circulate, observe whether the size of the numbers in the problem seems to be at the right level of challenge for each student. If not, you can easily adapt the problems by making the numbers larger or smaller.

How to Set Up the Choices   If you set up the activity choices at centers, show students what they will find at each one. Otherwise, make sure students know what materials they need and where to get them.

Choice 1: Paper Clip Problems—envelopes or resealable plastic bags with Paper Clip Problems, cubes, 300 charts, and calculators

Choice 2: Create Your Own Problems—cubes, 300 charts, calculators, drawing paper, and colored pencils, markers, or crayons

Choice 3: Related Problem Sets—envelopes or resealable plastic bags with Related Problem Sets, cubes, and 300 charts (no calculators for these problems)

Choice 1: Paper Clip Problems

Students choose one of the Paper Clip Problems and paste or glue it onto a piece of paper or into their mathematics notebook. They then work in pairs to solve the problem and write about how they solved it. If students are working with problems that are either too hard or too easy for them, help them choose a problem that is more appropriate.

❖ **Tip for the Linguistically Diverse Classroom**   Pair students of varying English proficiency. Those who are not writing comfortably in English can use a combination of numerals and drawings to tell how they solved the problems.

For students who have difficulty reading and interpreting the Paper Clip Problems, help them understand by drawing a quick sketch of the problem. For example:

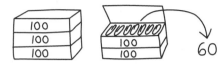

Model for students how they can draw their own sketches to help them visualize the problem.

Choice 2: Create Your Own Problems

Students create and solve their own problems patterned after the Paper Clip Problems. They write about something that was packaged in hundreds, then either add or subtract some amount. They can make their context realistic, silly, or strange, but must develop a good problem and solve it. Students write up and illustrate their problems, without the solutions, for others to solve.

❖ **Tip for the Linguistically Diverse Classroom** Again, pair students of varying English proficiency. Those proficient in English can write the problem while their partner illustrates it or provides drawings for key words.

Choice 3: Related Problem Sets

Students work on one or two problem sets. They paste or glue the problems they choose onto a piece of paper or into their mathematics notebook. These are problems like the ones they have been doing, but they are presented in standard notation. Students can solve the problems in any order, but should try to find ways they can use one problem in the group to help solve others. Tell students not to use calculators on these problems, because we want them to use strategies based on their number sense. You might have them check their problems with the calculator when they are finished and let you know which ones they had difficulty solving.

While Students Are Working on the Choices

Some time during Session 4, pause to have the whole class consider one or two Related Problem Sets. Ask them to explain how they would solve the problem and how they can use one of the problems to help with others in the cluster.

Sessions 3 and 4 Follow-Up

Problem Strategies   Students choose a Paper Clip problem and a Related Problem Set that they have not yet solved and glue or paste the problems onto Student Sheet 14, Problem Strategies. They will also need two copies of the 300 chart. Students take the materials home to work on.

 **Homework**

INVESTIGATION 5

Calendar Comparisons

What Happens

Session 1: How Much Longer? Students use calendars to figure out together how many more days until a particular holiday or event, and how many days until their next birthday. They work on an assessment activity in which they find distances between various time periods on the calendar.

Sessions 2 and 3: School Days During these two sessions, students consider two larger problems: (1) Which is more, the number of school days in a year or the number of non-school days? and (2) How much longer do children in other countries spend in school than we do? As an extension activity, students may write to legislators concerning their opinions about whether the school year should be extended, or do a survey of how others in their school feel about this issue.

Mathematical Emphasis

- Exploring the mathematical characteristics of the calendar and using it to add, subtract, and solve problems
- Solving complex problems that involve breaking the problem down into manageable parts
- Examining how parts and the whole are related in addition and subtraction (for example, that the number of school days and non-school days in a year should add up to a total of 365 days)

What to Plan Ahead of Time

Materials

- Calendars for the current year: 1 per pair (Sessions 1–3)
- School calendars for the current year: 1 per pair (Sessions 2–3)
- Wall calendar for display (Sessions 1–3, optional)
- Calculators (Sessions 2–3)
- Interlocking cubes and 300 charts remain available

Other Preparation

- Duplicate student sheets and teaching resources (located at the end of this unit) in the following quantities. If you have Student Activity Booklets, no copying is needed.

For Session 1

Student Sheet 15, Counting the Days (p. 100): 1 per student

For Sessions 2–3

School Days Around the World (p. 102): 1 per pair (classroom), 1 per student (homework)

Student Sheet 16, School Days Around the World Problems (p. 101): 1 per student (homework)

- Select two or three dates within the next month or so that have special meaning to the students. These could be the last day of school before a vacation, a holiday, a special celebration, or an upcoming school event. (Session 1)

How Much Longer?

What Happens

Students use calendars to figure out together how many more days until a particular holiday or event, and how many days until their next birthday. They work on an assessment activity in which they find distances between various time periods on the calendar. Students' work focuses on:

- doing comparisons that involve the question: How much longer?
- using a calendar to find and compare two dates

Materials

- Calendars for the current year (1 per pair)
- Student Sheet 15 (1 per student)
- Wall calendar (optional)
- Cubes and 300 charts

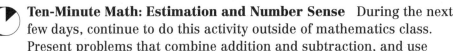

Ten-Minute Math: Estimation and Number Sense During the next few days, continue to do this activity outside of mathematics class. Present problems that combine addition and subtraction, and use some numbers in the hundreds. For example:

$$300 + 500 - 300 + 200 \qquad 600 - 300 + 400 - 200$$
$$500 - 2 + 500 + 3 \qquad 400 + 17 - 7 + 700$$

Remind students that looking over the whole problem first, reordering numbers to make the problem easier, and dealing with the largest numbers first are often good strategies. Students can use the calculator to double-check their mental arithmetic, and their mental arithmetic to double-check the calculator!

Activity

How Much Longer?

Distribute the calendars, one per pair, and ask everyone to find today's date. Talk about the date of an upcoming holiday or event that you've selected. Tell the class they will be figuring out how much longer it is until this important date, and discuss different ways of determining this.

Give students time to work this out with their partners. Students' strategies for counting the days will vary, and to some extent may depend on the date that you've chosen. Some will count by 1's, while others may use more complicated strategies. If you've chosen a date in the same calendar month, students may decide to subtract today's date from the target date. If the date is next month, they might count all the days until the end of this month, then add on the number of days next month until the target date. Some children may be able to make use of 7's as they count. Consider using a large wall calendar and asking a few students to talk about and demonstrate their strategies.

Expect some of the following questions to come up:

> Do we count today's date or not?
> Do we count the target date or not?
> Should we just count all the days in-between now and then?

These are important issues to talk about, because counting strategies really matter here. See the **Dialogue Box**, How Do You Count the Days? (p. 63), for an example of such a discussion.

After students have explained their different ways of counting, you may want to describe the social convention for counting days—that the length of time from today until tomorrow is one day, until the next day is two days, and so on. What we are really doing is counting the "jumps" from one day to the next. However, be sure to acknowledge that other methods students may use are perfectly reasonable, and that people could have decided to do it differently—and perhaps some people *have* done it differently in other places and times. Some students may notice that you get the same count by either counting the jumps, not counting today, or not counting the final day.

Write another important upcoming date on the board—a date that falls sometime in the next calendar month. If students have had trouble with the first date, make this one easier. If it seemed relatively easy, make this one harder. Ask students to work in pairs to figure out how much longer it is until that date. Write their answers on the board and be prepared for some differences. Again, encourage students to share their strategies.

How Many Days Until Your Birthday?

Students work in pairs to figure out how many more days it is until each of their next birthdays. There's considerable variation in the difficulty of this task. Students whose birthdays are coming up soon may have an easier time than those whose birthdays are six months from now. Students who just had birthdays may think that the task will be hardest for them. The idea of counting backwards and subtracting from 365 often doesn't occur to them.

Note: Some families do not celebrate individual birthdays. Be sensitive to this issue and, as needed, help students choose another date that is significant for them.

When students have finished, choose three or four different months to discuss—perhaps the month after this one, the month before this, and another about six months from now. Ask a few students who have birthdays in those months to share their strategies for finding out how much longer it is until their next birthday. You may want to have them use a wall calendar for showing how they worked.

Assessment

Counting the Days

Distribute Student Sheet 15, Counting the Days, and have students work on it individually for the rest of the session. They need their calendars to do this. If they do not finish, you may want them to complete the sheet as homework.

❖ **Tip for the Linguistically Diverse Classroom** Read the problems aloud. On problems 4 and 5, have students add arrows to help them remember whether the birthday is upcoming ($\rightarrow$) or just passed ($\leftarrow$).

As students work on this activity, walk around and see how they are using their calendars and how they are counting. Here are some questions to consider as you talk to students and as you examine their work later:

■ Do students know where to start and stop counting?

■ How are they counting? Do they have clear and efficient strategies? For example, do they break time into months and figure out the days in each month, then add them together? Do they rely solely on counting by 1's, even when counting the days in an entire month? In other words, are they counting up to 31 rather than knowing that they can add 31 to the number of days they are accumulating?

■ On the question about a birthday that is exactly 3 weeks from today, do they count the time by weeks? If not, do they know that 3 weeks = 21 days and count the days accurately?

■ Are they fluent in counting backward (for the birthday that happened recently) as well as forward?

■ When figuring out how long it is until August 3, do they combine the days from several months accurately? What kinds of combining strategies do they use and how well do these work?

How Do You Count the Days?

The following discussion took place in a classroom that was working on the first calendar activity, How Much Longer? (p. 60).

Everybody put an X on today, the 12th. OK. What's happening next Tuesday?

Michael: We're going on our field trip.

Right—that's the 17th. How many days from today is that?

Various students: Five . . . Six . . . Four.

How did you get your numbers?

Su-Mei: It's five because I counted from each day to the next day. From today till tomorrow is 1, then till Saturday is 2, till Sunday is 3, till Monday is 4, and till Tuesday is 5.

Annie: But today's a day, so we should count it. We should count the 12th, 13th, 14th, 15th, 16th, and 17th. That's six days.

Rashad: It's five days because today's not a full day. It already started.

Well, it's 9:30 now. Suppose we were starting our trip at 9:30 on Tuesday?

Yoshi: That would be four days.

How did you get that?

Yoshi: Tomorrow, Saturday, Sunday, Monday. So, four days.

Hmmm. So we have a lot of disagreement. Some people think we should count the day we start on. Some people don't. Some people think we should count from one day to the next as one day. What if your birthday were on Saturday? How many days away would you say it is?

Tamara: I'd say two, because there's almost all of today to get through, and then all of tomorrow, and then you wake up and it's your birthday.

Jeremy: Yes, it's two, but my reason is different. I'd say from today till tomorrow is one day, and

from tomorrow till Saturday is one day, so it's two days away.

Liliana: No, it's just one. I'd say it's one day away because today has already started and so you just count Friday.

Su-Mei: But then what if it was tomorrow? You don't say tomorrow is zero days away!

It's an interesting question, and one not easily answered. Most adults have decided that the way to count days is to count from one day to the next. So that tomorrow would be one day away, Saturday two days away, and Sunday three days away. If we use that method, how long is it from now until our field trip?

Mark: Five days. It really is closest to five days.

Annie: I still say it's six. It is if you count my way!

This conversation highlights an important mathematical issue—whether you are counting objects or intervals. These students are grappling with an important question: do you count days or the amount of time that elapses from one day to another? In this conversation, students are also beginning to think about what is discrete (such as individual objects) and what is continuous (such as measurements).

School Days

Materials

- School Days Around the World (1 per pair, classroom; 1 per student, homework)
- Calendars for the current year (1 per pair)
- School calendars (1 per pair)
- Cubes and 300 charts
- Calculators
- Wall calendar (optional)
- Student Sheet 16 (1 per student, homework)

What Happens

During these two sessions, students consider two larger problems: (1) Which is more, the number of school days in a year or the number of non-school days? and (2) How much longer do children in other countries spend in school than we do? As an extension activity, students may write to legislators concerning their opinions about whether the school year should be extended, or do a survey of how others in their school feel about this issue. Student work focuses on:

- using the calendar to compare two quantities
- considering the relationship between a whole (the calendar year) and its parts (school and non-school days)

Activity

Days In and Out of School

Which do you think there is more of in a year—the number of days that you're in school, or the number of days that you're not in school?

Distribute the yearly and school calendars and discuss students' ideas. Some students will think they're in school more because there are five school days and only two weekend days in a week. Others will notice vacation days and summer vacation. In your discussion, be sure to raise the issue of how many days are in an entire year, especially if this has not come up in Session 1. This will be important information in making the comparison.

Today, we're going to figure out which we really have more of each year—school days or non-school days. First, we have to agree on what days are not school days.

Go through the school calendar with the group. The class will need to do a little work to figure out which days are not school days. Students can participate in determining the following:

- how to count half-days, early release days, and other ambiguous days on your school's calendar
- which are the first and last days of school
- which days are school holidays
- which other days are non-school days, including summer vacation

Help students figure out a way of marking their calendars so that it is clear which days are school days and which are not. They could put an X on all non-school days, for example.

Now that we know which are the school days and which are the non-school days, you're going to work on these problems:

1. Which is more, the number of school days or the number of non-school days?

2. How many more?

Write these two questions on the board and have students work on them in pairs. Remind students to take notes to help them keep track of their counting and to double-check by using more than one strategy. When they are adding up their subtotals, they might use a calculator, but they should also make a mental estimate and check it by using cubes or a 300 chart, or adding them up in more than one way.

When students have finished, discuss their findings. Encourage several students to talk about how they got their numbers. You might want to begin this discussion by making a list on the board. Use the following headings to form three columns:

Number of   Number of   What's the
school days   non-school days   difference?

Write down the answers that students came up with in the appropriate columns. Don't expect everyone to have found the same totals. Ask students to tell which of the answers they agree with and why. In the discussion, some may point out that there are 365 days in a year, and that the school days and non-school days should add up to this number. Encourage this kind of thinking, and use it to help deal with discrepancies in the findings. Talk about how different ways of keeping track may have resulted in different numbers.

What methods for keeping track seemed to work the best? Which ones didn't work so well?

After some discussion, students may want to revise or double-check their work. For more information on the kinds of strategies students use, see the **Teacher Note,** Counting School and Non-School Days (p. 70).

When an agreement has been reached on the number of school and non-school days, go on to discuss how to figure out the difference. Ask if they were surprised about which was more, or if the size of the difference surprised them.

Activity

Making International Comparisons

Now we're going to figure out who goes to school the most days: students in this country or students in other countries. Do you think there are any differences?

As a class, discuss what those differences might be and why they might exist. If any of your students have been to school in other countries, this is a good opportunity for them to tell the class about the differences between the school years. For example, in many countries students go to school for at least half of Saturday.

Distribute copies of the chart, School Days Around the World, and ask students to examine it, noting that the information is current as of 1988. Use a world map to locate some of the countries named on the chart.

What do you notice on this chart? What surprises you about this information?

Pair up students and assign each pair one or two countries from the list. Students figure out how many more (or fewer) school days there are in those countries than in this country. Some students will have very easy comparisons. They can do a second, or even a third, if there is time.

After the students have had a few minutes to make the comparisons, have a representative from each pair write the difference between each country and the United States on the board. Ask some of them to describe how they figured out the differences.

Spend a few minutes discussing how many more days of school there are in the different countries. You might mention that there are many people in this country who think there should be more days of school each year. Use questions like the following to stimulate more discussion:

Would more days of school be a good idea? Why? What are some reasons that more days might be a bad idea?

Activity

What's Your Opinion?

After the class has had a chance to discuss the differences between the length of their own school year and those in other countries, raise this question:

Should children in this country go to school more days than they do now? What's your opinion? How long do you think the school year should be? Why?

Encourage students to use the comparisons they have made during this discussion. For example, they might point out how many additional days would be added to the current school year if a 200-day school year were approved, or how many fewer days of vacation they would get. They might also discuss what this would mean for their summer.

You may want to bring some of the following information into the discussion. People in favor of lengthening the school year to 200 days or more often give these reasons:

■ American schoolchildren need to have as much academic preparation as children from other countries, such as Japan.
■ There's more that children need to learn about now than they did a generation or two ago.
■ Over the summer, children often forget things that they have learned. This might not happen if they kept going to school in the summer.
■ In many families, all the adults work and they have a hard time arranging care for their children during summer vacation.
■ The major historical reason for school ending in May or June is that children used to help with farm chores, which were quite intensive in the spring and summer. This is no longer a consideration for most families.

The people opposed to this plan include a third grade girl in Massachusetts who has testified to a state committee that she wants to keep the length of the school year the way it is, because children need time off for recreation, and besides it would be too expensive to run schools all year long. The expense argument is one of the major reasons for opposing the plan. Here are other reasons:

- Buildings may not be air-conditioned in the summer, making conditions difficult for students and teachers alike.
- Teachers also get tired (it's a stressful job) and need the summer off to stay fresh.
- Teachers have other jobs in the summer, or take classes to further their own education.
- Families would not be able to take long vacations, and travel is educational, too.

❖ **Tip for the Linguistically Diverse Classroom** Have students write *More School Days?* at the top of a sheet of paper. They are to answer the question by writing yes or no, then support their opinion with a drawing that helps explain why. Give students the option of writing their opinion in their native language—both for this activity and for the letter in the extension activity. Pictures can also be sent with the letter.

Activity

Choosing Student Work to Save

As this unit ends, you may want to select one of the following options for creating a record of students' work on this unit.

- Students look back through their folders or notebooks and write about what they learned in this unit, what they remember most, and what was hard or easy for them. You might have students do this work during writing time.
- Students select one or two pieces of their work as their best work, and you also choose one or two pieces of their work, to be saved in a portfolio for the year. You might include students' recorded work on Student Sheet 12, Making Comparisons, and Student Sheet 15, Counting the Days, along with any other assessment tasks from this unit. Students can create a separate page with brief comments describing each piece of work saved.
- You may want to send a selection of students' work home for parents to see. Students write a cover letter, describing their work for this unit. This work should be returned if you are keeping a year-long portfolio of mathematics work for each student.

Sessions 2 and 3 Follow-Up

School Days Around the World Problems   Students solve the problems on Student Sheet 16, School Days Around the World Problems. They will also need a copy of School Days Around the World and a 300 chart.

Time Travel   Describe the system of time zones to students, using a globe or a flattened map of the world. Working in pairs, students answer questions such as:

What time is it in California when you start school in the morning?
What time is it in Australia when you get up? when you go to sleep?
When students in Tokyo are getting up, what are you doing?

Students can also make up questions for one another.

One or both of these extension activities could be done as part of your social studies, math, or writing class.

Writing About Your Opinion   After discussing the length of the school year, students may want to write about their opinions to one of their representatives or senators in Congress.

Taking a Survey   Students may be interested in surveying others, in their class or in their school, about their opinions on the issue of a longer school year. They might try to find out the following:

How many people think that the school year should be lengthened?
How many want it kept the way it is? or even shorter?
What are people's main reasons for their opinions?
Do third graders feel the same about this issue as their parents do?
What do teachers think?

Homework

Extensions

One of the major activities in Investigation 5 is counting and comparing the number of school and non-school days in a whole year. Solving this problem is a challenging task for many students. In classes that we have observed, students tend to use the following strategies:

Counting all the days one by one. Some students do not make use of units of time such as weeks or months. Instead, they choose a starting point and count each school day until they get to the end of the year. Then they do the same for the non-school days. Students who use this strategy occasionally write this information down at some point, but many rely on their memories alone. Many of them lose track. Work with students to help them find ways to count by larger quantities—by weeks or months or even just by 2's, 5's, or 10's—and to keep track of their counts.

Counting month by month. This is the most common strategy we have seen among third graders. Students separately count the school days and non-school days in each month and record this information. Once the counting is done, some students move immediately into adding the 12 monthly numbers in each category to get the total of school and non-school days. Other students get stuck at the point where they need to find the total. Occasionally, a student will figure out the difference between school and non-school days for each month, and keep track of the differences. This is a very sophisticated strategy.

Counting by weeks. This is a more elegant version of the first strategy. Rather than counting each single day, students deal with units of a week when it is appropriate (such as over the summer or during December vacation). They might determine that summer vacation is 11 weeks, for example, and multiply 11 by 7 in order to get the number of days in summer vacation.

Comparing the two numbers. The number of school and non-school days is actually quite close. There are typically 180 days of school, and 185 days when there is not school. Check to see how students compare these two numbers once they have found them. If students count on, check to see whether they end up with a correct number—or a number that is off by one in either direction. (If, with 180 and 185 days, they get a difference of 4, they are counting only the in-between days. If they get 6, they are counting all the numbers.) If a student moves immediately to the subtraction procedure, try to find an opportunity to ask why this procedure works.

Checking the findings. A few students check to make sure that the school days and non-school days add up to 365. A few even discover that once they know the number of school days, they don't need to count all of the non-school days. Some students think that using this subtraction shortcut is cheating. We think it's ingenious!

Exploring Data

Basic Activity

You or the students decide on something to observe about themselves. Because this is a Ten-Minute Math activity, the data they collect must be something they already know or can observe easily around them. Once the question is determined, quickly organize the data as students give individual answers to the question. The data can be organized as a line plot, a list, a table, or a bar graph. Then students describe what they can tell from the data, generate some new questions, and, if appropriate, make predictions about what will happen the next time they collect the same data.

Exploring Data is designed to give students many quick opportunities to collect, graph, describe, and interpret data about themselves and the world around them. Students focus on:

- describing important features of the data
- interpreting and posing questions about the data

Procedure

Step 1. Choose a question. Make sure the question involves data that students know or can observe: How many buttons are you wearing today? What month is your birthday? What is the best thing you ate yesterday? Are you wearing shoes or sneakers or sandals? How did you get to school today?

Step 2. Quickly collect and display the data. Use a list, a table, a line plot, or a bar graph. For example, a line plot for data about how many buttons students are wearing could look something like this:

Step 3. Ask students to describe the data. What do they notice about it? For data that have a numerical order (How many buttons do you have today? How many people live in your house? How many months until your birthday?), ask questions like these:

"Are the data spread out or close together? What is the highest and lowest value? Where do most of the data seem to fall? What seems typical or usual for this class?"

For data in categories (What is your favorite book? How do you get to school? What month is your birthday?), ask questions like these: "Which categories have a lot of data? few data? none? Is there a way to categorize the data differently to get other information?"

Step 4. Ask students to interpret and predict. "Why do you think that the data came out this way? Does anything about the data surprise you? Do you think we'd get similar data if we collected it tomorrow? next week? in another class? with adults?"

Step 5. List any new questions. Keep a running list of questions you can use for further data collection and analysis. You may want to ask some of these questions again.

Variations

Data from Home For homework, have students collect data that involves asking questions or making observations at home: What time do your brothers and sisters go to bed? What do you usually eat for breakfast?

Data from Another Class or Other Teachers Depending on your school situation, you may be able to assign students to collect data from other classrooms or other teachers. Students are always interested in surveying others about questions that interest them, such as this one: When you were little, did you like school?

Continued on next page

Categories If students take surveys about "favorites"—flavor of ice cream, breakfast cereal, book, color—or other data that fall into categories, the graphs are often flat and uninteresting. There is not too much to say, for example, about a graph like this:

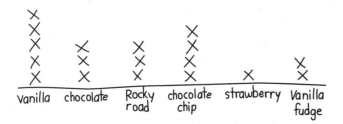

It is more interesting for students to group their results into more descriptive categories, so that they can see other things about the data. In this case, even though vanilla seems to be the favorite in the graph above, another way of grouping the data seems to show that flavors with some chocolate in them are really the favorites.

Chocolate flavors   ////   ////   //

Flavors without chocolate   ////   //

Estimation and Number Sense

Basic Activity

Students mentally estimate the answer to an arithmetic problem that they see displayed for about a minute. They discuss their estimates. Then they find a precise solution to the problem by using mental computation strategies.

Estimation and Number Sense provides opportunities for students to develop strategies for mental computation and for judging the reasonableness of the results of a computation done on paper or with a calculator. Students focus on:

- looking at a problem as a whole
- reordering or combining numbers within a problem for easier computation
- looking at the largest part of each number first (looking at hundreds before tens, thousands before hundreds, and so forth)

Materials

Calculators (for variation)

Procedure

Step 1. Present a problem on the chalkboard or overhead. For example:

$9 + 62 + 91 + 30$

Step 2. Allow 15–20 seconds for students to think about the problem. In this time, students come up with the best estimate they can for the solution. This solution might be—but does not have to be—an exact answer. Students do not write anything down or use the calculator during this time.

Step 3. Cover the problem and ask students to discuss what they know. Ask questions like these: "What did you notice about the numbers in this problem? Did you estimate an answer? How did you make your estimate?"

Encourage all kinds of statements and strategies. Some will be estimates; others may be quite precise:

> "It's definitely bigger than 100 because I saw a 90 and a 60."

> "It has to be 192 because the 91 and the 9 make 100 and the 30 and the 62 make 92."

Be sure that you continue to encourage a variety of observations, especially the "more than, less than" statements, even if some students have solved it exactly.

Step 4. Uncover the problem and continue the discussion. Ask further: "What do you notice now? What do you think about your estimates? Do you want to change them? What are some mental strategies you can use to solve the problem exactly?"

Variations

Problems That Can Be Reordered Give problems like the following examples, in which grouping the numbers in particular ways can help solve the problem easily:

$6 + 2 - 4 + 1 - 5 + 4 + 5 - 2$

$36 + 22 + 4 + 8$

$112 - 30 + 60 - 2$

$654 - 12 + 300 + 112$

Encourage students to look at the problem as a whole before they start to solve it. Rather than using each number and operation in sequence, they see what numbers are easy to put together to give answers to part of the problem. Then they combine their partial results to solve the whole problem.

Problems with Large Numbers Present problems that require students to "think from left to right" and to round numbers to "nice numbers" in order to come up with a good estimate. For example:

$230 + 343 + 692$	$\$3.15 \times 9$
$8 + 1200 + 130$	$\$5.13$
	$\$6.50$
	$+ \$3.30$

Present problems in both horizontal and vertical formats. If the vertical format triggers a rote

Continued on next page

procedure of starting from the right and "carry-ing," encourage students to look at the numbers as a whole, and to think about the largest parts of the numbers first. Thus, for the problem 230 + 343 + 692, they might think first, "About how much is 692?—700." Then, thinking in terms of the largest part of the numbers first (hundreds), they might reason: "300 and 700 is a thousand, and 200 more is 1200, and then there's some extra, so I think it's a little over 1200."

Is It Bigger or Smaller? Use any of the kinds of problems suggested above, but pose a question about the result to help students focus their estimation: "Is this bigger than 20? Is it smaller than $10.00? If I have $20.00, do I have enough to buy these four things?"

Using the Calculator The calculator can be used to check results. Emphasize that it is easy to make mistakes on a calculator, and that many people who use calculators all the time often make mistakes. Sometimes you punch in the wrong key or the wrong operation. Sometimes you leave out a number by accident, or a key sticks on the calculator and doesn't register. However, people who are good at using the calculator always make a mental estimate so they can tell whether their result is reasonable.

Pose some problems like this one:

> I was adding 212, 357, and 436 on my calculator. The answer I got was 615. Was that a reasonable answer? Why do you think so?

Include problems in which the result *is* reasonable and problems in which it is *not*. When the answer is unreasonable, some students might be interested in figuring out what happened. For example, in the above case, I accidentally punched in 46 instead of 436.

Related Homework Options

Problems with Many Numbers Give one problem with many numbers that must be added and subtracted. Students show how they can reorder the numbers in the problem to make it easier to solve. They solve the problem using two different methods to double-check their solution. One way might be using the calculator. Here is an example of such a problem:

$$30 - 6 + 92 - 20 + 56 + 70 + 8$$

The following activities will help ensure that this unit is comprehensible to students who are acquiring English as a second language. The suggested approach is based on *The Natural Approach: Language Acquisition in the Classroom* by Stephen D. Krashen and Tracy D. Terrell (Alemany Press, 1983). The intent is for second-language learners to acquire new vocabulary in an active, meaningful context.

Note that *acquiring* a word is different from *learning* a word. Depending on their level of proficiency, students may be able to comprehend a word upon hearing it during an investigation, without being able to say it. Other students may be able to use the word orally, but not read or write it. The goal is to help students naturally acquire targeted vocabulary at their present level of proficiency.

We suggest using these activities just before the related investigations. The activities can also be led by English-proficient students.

Investigation 1

record, tie

1. Show students *The Guinness Book of Records,* and explain that this book tells who set a record by doing something better, longer, or faster than anyone else. Some of the records set have been very unusual. As you give some examples from the book, ask students to predict the record set for each category. (For example: This page shows the record set for the largest fish ever caught with a fishing rod. How large do you think this fish was? This page shows the record number of days that someone sneezed. How many days do you think this was?) Write the records on the board.

2. Tell students that if another person does something just as well, as long, or as fast as the record holder, then that person ties the record.

3. As you point to each record on the board, pretend to be a radio announcer broadcasting how these records have now been tied by students in your class. (For example: Just yesterday, Liliana tied the world record for catching the largest fish. The record holder, Alf Dean, caught a great white shark that weighed 2,664 pounds—and now so has Liliana! Now for another news

flash—Dylan has just tied the world record for the longest time anyone has sneezed. Just like the record holder, Donna Griffiths, Dylan has sneezed for 977 days!)

Investigation 2

weight, balance

1. Make a sketch of a seesaw on the board. Draw a stick figure sitting on one end of the seesaw and two figures standing beside it.

2. Sketch a bathroom scale.

3. Point to the figure sitting on the seesaw and then to the scale. Tell students that this person's weight is 65 pounds. Write 65 under the figure. Tell and write the weight of the two other figures: 128 pounds and 65 pounds.

4. Ask students what would happen if the person weighing 128 pounds sat on the other end of the seesaw. Have a student draw what the seesaw would look like.

5. Point to the original drawing of the seesaw and ask what would happen if the 65-pound person sat on the other end. Ask a student to draw this seesaw.

6. Explain that the seesaw has the same weight on each side and is now balanced.

Investigation 3

coupon

1. Give each student a grocery coupon with a picture of the item on it. Ask students if anyone in their family uses coupons.

2. Explain that when someone uses a coupon in a store, the item costs less. Give an example: A can of beans costs 65 cents. Dominic has a coupon that will save him 25 cents; so, with the coupon, the beans will cost 40 cents.

3. Ask several students to tell what item is shown on their coupons and how much money they will save if they buy that item in a store.

hours, minutes

1. Draw a clock on the board that shows 5 o'clock. Ask students what time the clock says. Then show the clock at 5:10. Ask students what time it is now and how many minutes have passed since 5 o'clock.

2. Tell students that at 5:10 tonight you are going to make a salad. It will take you ten minutes to make the salad. What time will the clock say when you have finished making the salad? Have a volunteer draw the clock at 5:20.

3. Have students suggest activities and the time it would take to do them. As each activity is suggested, have a student redraw the clock to tell what time it will be when the activity is finished.

4. When students get to 6 o'clock, have them stop and note that an hour has passed since 5 o'clock.

birthday

1. Pantomime holding a newborn baby in your arms. Tell students that the baby was just born, and write today's date on the board. Explain that this is the baby's birthday.

2. Sing "Happy Birthday," and encourage students to sing along with you. Ask for volunteers to sing the song in their native languages. Students might also make a poster with the words "Happy Birthday" in all the languages that are spoken in the class.

3. Have each student point to his or her birthday on a calendar. This is also a good opportunity to make a graph of the months of students' birthdays.

Blackline Masters

Dear Family,

During the next few weeks in math, our class will be working on an addition and subtraction unit called *Combining and Comparing*. We'll be focusing on interesting and real problems that involve combining and comparing two or more amounts.

Students will be learning about many different ways to solve addition and subtraction problems. Students will be encouraged to develop more than one way to solve a problem and to use methods that are based on understanding numbers and their relationships. Some of these methods may not be the ones you learned in school, but you may recognize some of them as methods you use in your daily life. We encourage students to develop strategies that make sense to them, that they can use easily and flexibly. For example, one of the first things we will do is compare the age of your family's oldest relative with the age of the oldest person on record (120 years old). Suppose your oldest relative is 83 years old. Your child might find the answer by seeing how many years you must add on to get from 83 to 120. Add 7 to 83 to get to 90, 10 more to get to 100, 20 more to get to 120. Another approach could be to subtract 80 from 120 to get 40, then subtract 3 more to get 37.

The point is for children to find useful and meaningful ways of putting together and taking apart numbers. While our class is working on this unit, you can help in several ways:

> When your child has an assignment to do at home—such as collecting data about the ages of pets and oldest relatives—offer your help, and ask your child about what he or she is doing in class.

> Ask your child to describe any of the homework problems and tell you about the strategy used to solve it. Communication is an important part of mathematics, and students need to describe their strategies through talking, writing, drawing, or using concrete objects. You can be an important audience.

> You can also share your own ideas. At one point, we will work on the mathematics of "party planning." You might explain how you would figure out how to fit a number of different activities into a two-hour block of time.

An important emphasis in this unit is for students to recognize when and how to apply addition and subtraction and to develop procedures for adding and subtracting that they understand thoroughly and can use confidently. One of the most important things you can do is to show genuine interest in the ways your child solves problems, even if they are different from your own.

Thanks for your help and your interest in your child's mathematics.

Sincerely,

Close to 100 Score Sheet

Game 1   Score

Round 1:   ___ ___ + ___ ___ = _____   _____

Round 2:   ___ ___ + ___ ___ = _____   _____

Round 3:   ___ ___ + ___ ___ = _____   _____

Round 4:   ___ ___ + ___ ___ = _____   _____

Round 5:   ___ ___ + ___ ___ = _____   _____

TOTAL SCORE   _____

Game 2   Score

Round 1:   ___ ___ + ___ ___ = _____   _____

Round 2:   ___ ___ + ___ ___ = _____   _____

Round 3:   ___ ___ + ___ ___ = _____   _____

Round 4:   ___ ___ + ___ ___ = _____   _____

Round 5:   ___ ___ + ___ ___ = _____   _____

TOTAL SCORE   _____

                       Investigation 1 • Sessions 1–2
Combining and Comparing

Ages

1. **Pets**   Fill in this chart with information about pets you know. The pets can belong to your family or to someone else.

Name of Pet	Type of Animal	Age
_____	_____	_____
_____	_____	_____
_____	_____	_____
_____	_____	_____

Record Ages for Animals

Guinea pig	14 years	
Dog	29 years	
Cat	36 years	
Gerbil	8 years	
Mouse	7 years	
Rabbit	18 years	

Do you have one of these animals on your list?

What is the difference between your pet's age and the record age?

Show your work on the back of this sheet.

2. **Oldest Relative**   Talk to your family about who your oldest living relative might be.

Name of person _____ Age _____

How is this person related to you?_____

1	2	3	4	5	6	7	8	9	10
11	12	13	14	15	16	17	18	19	20
21	22	23	24	25	26	27	28	29	30
31	32	33	34	35	36	37	38	39	40
41	42	43	44	45	46	47	48	49	50
51	52	53	54	55	56	57	58	59	60
61	62	63	64	65	66	67	68	69	70
71	72	73	74	75	76	77	78	79	80
81	82	83	84	85	86	87	88	89	90
91	92	93	94	95	96	97	98	99	100

Weights and Comparisons

Kind of Food	Starting Weight	Second Weight	Difference

Investigation 2 • Session 1
Combining and Comparing

Who Is Closer to 100?

Two students are playing Close to 100.

Student 1 has   Student 2 has

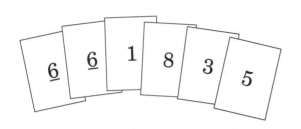   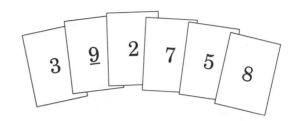

Find the 4 cards that will get each player
as close as possible to 100.

Student 1   Student 2

☐☐ + ☐☐ = _____   ☐☐ + ☐☐ = _____

Who got closer to 100? _____

Choose one student's hand. Explain why this
is as close as you can get to 100 with those six cards.

_____

_____

_____

_____

_____

_____

_____

Children's Heights

Part 1

Six third graders measured each other's heights.
They found out that the total of all six heights
was 318 inches.

There are lots of ways the six heights can
total 318 inches. Show one way by filling in
the chart below. Make it realistic.

Student 1 is _____ inches tall.	Student 4 is _____ inches tall.
Student 2 is _____ inches tall.	Student 5 is _____ inches tall.
Student 3 is _____ inches tall.	Student 6 is _____ inches tall.
The total is 318 inches.	

Part 2

In the chart below, fill in your own height.
Then fill in your partner's height.
Then fill in the rest of the heights
so the total is still 318 inches.

I am _____ inches tall.	Student 4 is _____ inches tall.
My partner is _____ inches tall.	Student 5 is _____ inches tall.
Student 3 is _____ inches tall.	Student 6 is _____ inches tall.
The total is 318 inches.	

On the back of this sheet, show how you reached both of your
solutions. Use + and = signs.

Coupons Add Up

Get ready to go shopping. Use the sheets of coupons.

1. Find and list coupons that add up to $2.50.

_____

_____

_____

2. Find and list coupons that add up to $3.70.
   Do not use cereal coupons.

_____

_____

_____

3. Find and list coupons that add up to $3.25.
   Use only coupons for foods you would like to eat.

_____

_____

_____

4. Find and list coupons that add up to $2.75.
   Use only coupons for things you cannot eat.

_____

_____

_____

Different Ways to Add

Solve each problem three different ways.
Using a calculator can be one way.
Make notes about how you solved the problems.
Be sure that others can understand what you did.

---

1. $42 + 36 + 18 =$

First way:

Second way:

Third way:

---

2. $57 + 3 + 56 =$

First way:

Second way:

Third way:

---

3. $125 + 53 + 27 =$

First way:

Second way:

Third way:

Planning a Party

1. On another piece of paper, write down all the things you would like to do at your party. You might want to play games, eat, listen to stories, or play outside.

2. Now make a schedule for your party. The activities need to fit from 5:00 P.M. to 7:00 P.M. Your friends have to go home right at 7 o'clock!

   ■ How many minutes will each activity take?

   ■ When will each activity start?

Party activity	Starting time	How many minutes?

How many minutes does your party last altogether? _____

Choose a Problem

Choose two or three of these problems.
Solve each one at least two different ways.
Show your work so that you can explain what you did.

1. $26 + 27 + 25 =$	**2.** $55 + 18 + 45 =$
3. $45 + 25 + 13 =$	**4.** $16 + 73 + 4 =$
5. $52 + 99 =$	**6.** $23 + 14 + 37 + 5 + 26 =$

$1.25

Crunchy Corn Cereal

$ .70

Fruity Fig Cereal

$1.00

Soft Sponges

$ .45

Frozen Fish Sticks

$ .40

Happy Apple Juice

25¢

Cloudy Cleanser

40¢

Minty Toothpaste

80¢

Pumpkin Pudding Surprise

69¢

Super Soap

20¢

Snack-Time Raisins

Investigation 3 • Resource
Combining and Comparing

$1.75 Low-Salt Soy Sauce	50¢ Wipe Up Paper Towels
$ .25 Carmen's Tacos	31¢ Perfect Pencils
15¢ Trusty Trash Bags	$ .60 Ice Cold Frozen Yogurt
$ .75 Tom's Corn Bread Mix	30¢ Wheat Yum Crackers
$1.20 Color Bright Film	$1.80 Nutty Crunch Cereal

–23	–24	–25	–26	–27	–28	–29	–30	–31	–32
–22									
–21	–20	–19	–18	–17	–16	–15	–14	–13	–12
									–11
–1	–2	–3	–4	–5	–6	–7	–8	–9	–10
0									
+1	+2	+3	+4	+5	+6	+7	+8	+9	+10
									+11
+21	+20	+19	+18	+17	+16	+15	+14	+13	+12
+22									
+23	+24	+25	+26	+27	+28	+29	+30	+31	+32

Handfuls at Home

Find something around your house that you can grab by handfuls. Compare the handfuls for your left hand and right hand.

Ask other people in your family to try this too. Record your data below.

We took handfuls of _____ .

Draw a picture of what you grabbed, or glue some examples of it here:

Person's name	Number in right hand	Number in left hand	Difference

Handfuls of Beans

1. Suppose you can hold 189 beans in your right hand and 150 beans in your left hand.
   How many more beans are in your right hand? _____

   Write down how you figured this out.

2. Suppose you can hold 150 beans in your right hand and 217 beans in your left hand.
   How many more beans are in your left hand? _____

   Write down how you figured this out.

Making Comparisons

Look at the class data for handfuls of beans.
Use it to fill in the numbers below.

Class Data

Largest **right** handful: _____   Largest **left** handful: _____

Smallest **right** handful: _____   Smallest **left** handful: _____

1. What is the difference between the largest right handful
   and the smallest right handful?

   How I solved it:

   Writing the problem with + and – signs:

   _____ + _____ = _____   _____ – _____ = _____

2. What is the difference between the largest left handful
   and the smallest left handful?

   How I solved it:

   Writing the problem with + and – signs:

   _____ + _____ = _____   _____ – _____ = _____

Class Handfuls and My Handfuls

Look at the class data for handfuls of beans.
Use it to fill in the numbers below.

Class Data

Largest **right** handful: _____   Largest **left** handful: _____

Smallest **right** handful: _____   Smallest **left** handful: _____

My Data

My **right** handful: _____   My **left** handful: _____

1. What is the difference between your right handful
   and the largest right handful in the class?

   How I solved it:

   Writing the problem with + and − signs:

   _____ + _____ = _____   _____ − _____ = _____

2. What is the difference between your left handful
   and the smallest left handful in the class?
   How I solved it:

   Writing the problem with + and − signs:

   _____ + _____ = _____   _____ − _____ = _____

Problem Strategies

1. The problem I am solving is:

   This is how I did it:

2. The problem I am solving is:

   This is how I did it:

QUICK IMAGE PAPER CLIP BOXES

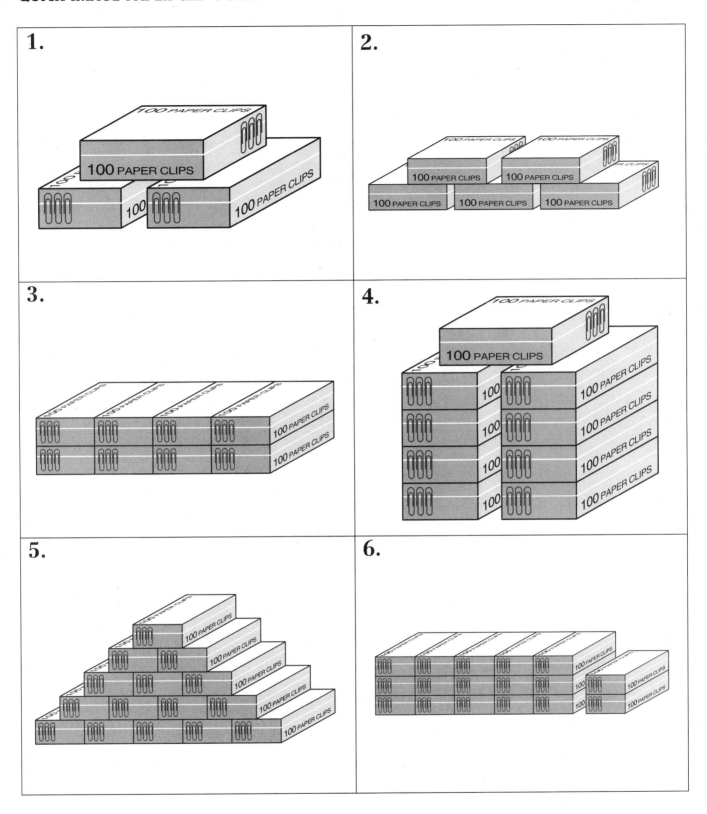

We had 500 paper clips. We used 60 clips to make a chain.

How many paper clips are left?

We had 9 boxes of paper clips. Then 10 paper clips fell out of one box.

How many paper clips are left in the boxes?

1. There were 250 paper clips on the shelf. Someone put 300 more on the shelf. How many are there now?

2. After a while, someone put 15 more paper clips on the shelf. Now how many are there?

1. Oops! The factory put 101 paper clips in each box. How many paper clips will you get if you buy 9 of these boxes?

2. What if they put 99 clips in each box? How many will there be in 5 boxes?

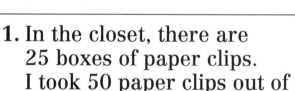

1. We had 1100 paper clips. We used 40 of them. How many paper clips are left?

2. Oops! We need 8 more clips, so we take those out of a box. Now how many clips are left?

1. In the closet, there are 25 boxes of paper clips. I took 50 paper clips out of one box. How many paper clips are left in the closet?

2. Someone took 5 more clips out of that box. How many are left now?

98

Problem Set 1

$100 - 5 =$

$200 - 5 =$

$300 - 5 =$

Problem Set 2

$300 - 5 =$

$301 - 5 =$

$301 - 4 =$

Problem Set 3

$$\begin{array}{r} 200 \\ 200 \\ + 200 \\ \hline \end{array} \qquad \begin{array}{r} 200 \\ 199 \\ + 199 \\ \hline \end{array} \qquad \begin{array}{r} 201 \\ 200 \\ + 199 \\ \hline \end{array}$$

Problem Set 4

$$\begin{array}{r} 300 \\ - 50 \\ \hline \end{array} \qquad \begin{array}{r} 300 \\ - 55 \\ \hline \end{array} \qquad \begin{array}{r} 301 \\ - 50 \\ \hline \end{array}$$

Problem Set 5

$500 - 10 =$

$600 - 20 =$

$700 - 30 =$

Problem Set 6

$500 - 30 =$

$600 - 30 =$

$700 - 30 =$

$1000 - 30 =$

Problem Set 7

$300 + 100 + 600 - 1 =$

$300 + 100 - 9 + 600 - 1 =$

$300 - 9 + 600 + 10 - 1 =$

Investigation 4 • Resource
Combining and Comparing

Counting the Days

1. What is today's date? _____

2. Cesar's birthday is April 30.
   How long is it until his birthday? _____
   Show how you figured it out.

3. Kate's birthday is August 3.
   How long is it until her birthday? _____
   Show how you figured it out.

4. Seung's birthday is coming up
   exactly 3 weeks from today.
   What is the date of her birthday? _____

5. Dylan's birthday was 13 days ago.
   What is the date of his birthday? _____

School Days Around the World Problems

Use the School Days Around the World sheet to solve these problems.

1. _____ is a country where
   A

   students go to school the fewest days of the countries on this chart.

   _____ is a country where
   B

   students go to school the most days of the countries on this chart.

   How many more days do the students in country B go to school than the students in country A? Show how you figured this out.

2. In _____ students go to
   A

   school for 200 days.

   In _____ students go to
   B

   school less than 200 days.

   How many more days do the students in country A go to school than the students in country B? Show how you figured this out.

Number of School Days in a Year

Country	School Days	Country	School Days
Algeria	200	Ireland	184
Argentina	184	Israel	216
Australia	220	Jamaica	190
Belgium	180	Japan	243
Bolivia	160	Jordan	184
Botswana	200	Luxembourg	216
Brazil	200	Mexico	200
Cambodia	190	Morocco	175
China	200	Nigeria	190
Dominican Republic	190	Portugal	180
England	192	Russia	211
Finland	191	South Africa	174
Flemish Belgium	160	South Korea	220
France	185	Spain	180
Germany	240	Swaziland	191
Guatemala	200	Thailand	200
Guinea	193	United States	180
Haiti	175	Zambia	180
Hong Kong	220	Zimbabwe	225

[Source: Data from George Thomas Kurian, editor. *World Education Encyclopedia.*
New York: Facts on File, 1988.]

0	0	1	1
0	0	1	1
2	2	3	3
2	2	3	3

Unit Resource
Combining and Comparing

4	4	5	5
4	4	5	5
<u>6</u>	<u>6</u>	7	7
<u>6</u>	<u>6</u>	7	7

8	8	9	9
8	8	9	9

WILD CARD	WILD CARD
WILD CARD	WILD CARD

Unit Resource
Combining and Comparing

Materials

- One deck of Numeral Cards
- Close to 100 Score Sheet for each player

Players: 1, 2, or 3

How to Play

1. Deal out six Numeral Cards to each player.

2. Use any four of your cards to make two numbers. For example: 6 and 5 could make either 56 or 65.

   Wild Cards can be used as any numeral.

   Try to make numbers that, when added, give you a total that is close to 100.

3. Write these two numbers and their total on the Close to 100 Score Sheet. For example: 42 + 56 = 98.

4. Find your score. Your score is the difference between your total and 100. For example: If your total is 98, your score is 2. If your total is 105, your score is 5.

5. Put the cards you used in a discard pile. Keep the two cards you didn't use for the next round.

6. For the next round, deal four new cards to each player. Make more numbers that come close to 100. When you run out of cards, mix up the discard pile and use them again.

7. Five rounds make one game. Total your scores for the five rounds. LOWEST score wins!

300 CHART

1	2	3	4	5	6	7	8	9	10
11	12	13	14	15	16	17	18	19	20
21	22	23	24	25	26	27	28	29	30
31	32	33	34	35	36	37	38	39	40
41	42	43	44	45	46	47	48	49	50
51	52	53	54	55	56	57	58	59	60
61	62	63	64	65	66	67	68	69	70
71	72	73	74	75	76	77	78	79	80
81	82	83	84	85	86	87	88	89	90
91	92	93	94	95	96	97	98	99	100
101	102	103	104	105	106	107	108	109	110
111	112	113	114	115	116	117	118	119	120
121	122	123	124	125	126	127	128	129	130
131	132	133	134	135	136	137	138	139	140
141	142	143	144	145	146	147	148	149	150
151	152	153	154	155	156	157	158	159	160
161	162	163	164	165	166	167	168	169	170
171	172	173	174	175	176	177	178	179	180
181	182	183	184	185	186	187	188	189	190
191	192	193	194	195	196	197	198	199	200
201	202	203	204	205	206	207	208	209	210
211	212	213	214	215	216	217	218	219	220
221	222	223	224	225	226	227	228	229	230
231	232	233	234	235	236	237	238	239	240
241	242	243	244	245	246	247	248	249	250
251	252	253	254	255	256	257	258	259	260
261	262	263	264	265	266	267	268	269	270
271	272	273	274	275	276	277	278	279	280
281	282	283	284	285	286	287	288	289	290
291	292	293	294	295	296	297	298	299	300

Practice Pages

This optional section provides homework ideas for teachers who want or need to give more homework than is assigned to accompany the activities in this unit. The problems included here provide additional practice in learning about number relationships and in solving computation and number problems. For number units, you may want to use some of these if your students need more work in these areas or if you want to assign daily homework. For other units, you can use these problems so that students can continue to work on developing number and computation sense while they are focusing on other mathematical content in class. We recommend that you introduce activities in class before assigning related problems for homework.

The Arranging Chairs Puzzle   This activity is introduced in the unit *Things That Come in Groups*. If your students are familiar with the activity, you can simply send home the directions so that students can play at home. If your students have not done this activity before, introduce it in class and have students do it once or twice before sending it home. Early in the year, ask students to work with numbers such as 15, 18, and 24. Later in the year, as they become ready to work with larger numbers, they can try numbers such as 32, 42, or 50. You might have students do this activity two times for homework in this unit.

Money Problems   This type of problem is introduced in the unit *Mathematical Thinking at Grade 3*. Here you are provided three of these problems for student homework. You can make up other problems in this format, using numbers that are appropriate for your students. Students record their strategies for solving the problems, using numbers, words, or pictures.

Addition in Two Ways   Solving problems in two ways is emphasized throughout this level of *Investigations*. Here we provide three sheets of addition problems that students solve in two different ways. You can make up other problems in this format, using numbers that are appropriate for your students. Students describe each way they solved the problem. We recommend that you give students an opportunity to share a variety of strategies for solving addition problems before you assign this homework.

The Arranging Chairs Puzzle

What You Will Need

30 small objects to use as chairs (for example, cubes, blocks, tiles, chips, pennies, buttons)

What to Do

1. Choose a number greater than 30.

2. Figure out all the ways you can arrange that many chairs. Each row must have the same number of chairs. Your arrangements will make rectangles of different sizes.

3. Write down the dimensions of each rectangle you make.

4. Choose another number and start again. Be sure to make a new list of dimensions for each new number.

Example
All the ways to arrange
12 chairs

Dimensions
1 by 12
12 by 1
2 by 6
6 by 2
3 by 4
4 by 3

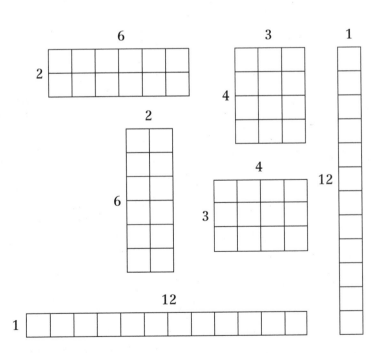

Practice Page A

I picked peaches from our peach tree. My mother paid me 2¢ for each peach I picked. I picked 52 peaches. How much did I earn?

Show how you solved this problem. You can use numbers, words, or pictures.

Practice Page B

Two friends earned 55¢. They want to split it evenly.
How much should each one get?

Show how you solved this problem. You can use
numbers, words, or pictures.

Practice Page C

I have 96¢. I put half the money in one pocket and half the money in the other pocket. How much money did I put in one pocket?

Show how you solved this problem. You can use numbers, words, or pictures.

Practice Page D

Solve this problem in two different ways, and write about how you solved it:

$$108 + 39 =$$

Here is the first way I solved it:

Here is the second way I solved it:

Practice Page E

Solve this problem in two different ways, and write about how you solved it:

192 + 103 =

Here is the first way I solved it:

Here is the second way I solved it:

Practice Page F

Solve this problem in two different ways, and write about how you solved it:

$$234 + 123 =$$

Here is the first way I solved it:

Here is the second way I solved it: